Gyroscopes - Principles and Applications

Edited by Xuye Zhuang and Lianqun Zhou

Published in London, United Kingdom

IntechOpen

Supporting open minds since 2005

Gyroscopes - Principles and Applications
http://dx.doi.org/10.5772/intechopen.78870
Edited by Xuye Zhuang and Lianqun Zhou

Contributors
Hongyu Zhao, Sen Qiu, Zhelong Wang, Ning Yang, Jie Li, Jianjun Wang, Mundla Narasimhappa, Ramón José Pérez Menéndez, Wenlong Han, Xuejuan Lin, Ke Chen, Wen Zhang, Yuichi Ikeda, Xuye Zhuang, Dongxing Li, Pinghua Li, Wentao Sui

First published in London, United Kingdom, 2020 by IntechOpen
IntechOpen is the global imprint of INTECHOPEN LIMITED, registered in England and Wales, registration number: 11086078, 7th floor, 10 Lower Thames Street, London, EC3R 6AF, United Kingdom
Printed in Croatia

British Library Cataloguing-in-Publication Data
A catalogue record for this book is available from the British Library

Additional hard and PDF copies can be obtained from orders@intechopen.com

Gyroscopes - Principles and Applications
Edited by Xuye Zhuang and Lianqun Zhou
p. cm.
Print ISBN 978-1-78985-057-4
Online ISBN 978-1-78985-058-1
eBook (PDF) ISBN 978-1-83880-390-2

For EU product safety concerns:
IN TECH d.o.o., Prolaz Marije Krucifikse Kozulić 3, 51000 Rijeka, Croatia, info@intechopen.com or visit our website at intechopen.com.

Meet the editors

Xuye Zhuang received his BS degree in Mechanical Engineering in 2004 from the Southwest Petroleum University. Since 2004 he studied at the Changchun Institute of Optics, Fine Mechanics and Physics, Chinese Academy of Sciences, and received his PhD degree in Engineering Science from the Graduate School of the Chinese Academy of Sciences (currently named University of Chinese Academy of Sciences) in 2009. He is currently working as a full professor in the School of Mechanical Engineering, Shandong University of Technology. He has been appointed a member of Young Taishan Scholars of Shandong Province, China. His research interest focuses on micro/nanosensors and MEMS/NEMS. He has about 20 patents and published more than 30 articles in international journals and conferences.

Lianqun Zhou studied at the Changchun Institute of Optics, Fine Mechanics and Physics, Chinese Academy of Sciences from 2003. Since 2007, he studied at the Micro Nano Sciences & Systems Department (MN2S) of the FEMTO-ST Institute, Besançon, France (www.femto-st.fr). He received his PhD degree in Engineering Science in 2010 at the University of Franche-Comté, Besançon, France, and the Graduate School of the Chinese Academy of Sciences (now named University of Chinese Academy of Sciences). He is currently working as a full professor at the Suzhou Institute of Biomedical Engineering and Technology, Chinese Academy of Sciences. His research interest focuses on the development of nanomaterials, micro/nanosensors, and MEMS chips. He has published more than 30 articles in international journals and is the coauthor of 20 patents.

Contents

Preface

Gyroscopes have been part of our lives for a long time. They have been used as toys in our childhood, and as navigation tools to equip our spacecraft, aircraft, vehicles, vessels, and even our smartphones to make our lives safe and comfortable. Stone tops have been excavated from the Neolithic sites in Xia County, Shanxi, China. It can be seen that gyroscopes have a history of at least four or five thousand years. A top belongs to the mechanical gyroscope, which is the most common or familiar type of gyroscope. A mechanical gyroscope spins on a point when it is turned around very quickly. However, the name "gyroscope" did not appear until in the middle of the nineteenth century. It was created by a French physicist, Jean-Bernard-Léon Foucault, by joining two Greek roots: gyros meaning "circle or rotation" and skopeein meaning "to see." Since then the field of gyroscopes has been maintaining a momentum of vigorous development and expansion, influenced by new applications of the latest scientific and technological innovations. New types of gyroscopes and new applications are springing up like mushrooms.

This book reviews recent topics on gyroscopes. Chapter 1 briefly introduces the history of gyroscopes, and presents a concise analysis of four main types of gyroscope: mechanical gyroscope, ring laser gyroscope, fiber-optic gyroscope (FOG), and MEMS (microelectromechanical systems) gyroscope. The dynamic future of new gyroscopes based on new principles and technologies is also presented.

Chapter 2 analyzes the classical structure and main performance parameters of the interferometric fiber-optic gyroscope (IFOG) and the integrated optics passive-resonator gyroscope (IORG). The main advanced models and performance parameters of these two types of inertial sensors are described and the design trends of both types are forecast. The chapter demonstrates that IFOGs have higher resolution performance than resonant fiber-optic gyroscopes and IORGs. IORG technology has experienced a vigorous development and refinement, and yet its performance is still at least one order of magnitude worse than that demanded by navigation applications. An improvement in this kind of gyroscope is needed to realize a significant impact on the market.

Chapter 3 reviews the developmental progress of FOGs, and also introduces their basic principles and application areas. The authors analyze the characteristics of the three classical types of FOGs: interferometric FOGs, resonant FOGs, and stimulated Brillouin scattering FOGs. The chapter presents a comparative analysis of the development and research situation of FOGs in the United States, Japan, France, and other major developing countries, and compares the application of FOGs in various international companies. The developmental trends and key technological breakthroughs of FOGs are also forecast.

In Chapter 4, low-cost MEMS gyroscope noise behavior is characterized using an ARMA autoregressive-moving-average (ARMA) model. A linear Sage Husa adaptive fading Kalman filter based on an ARMA (2, 1) model with adaptive transitive factors is introduced to reduce the drift and random noise of MEMS gyroscopes.

The proposed method consists of two stages. In the first stage, the predicted state vector is modified by an adaptive transitive factor. In the second stage, the measurement noise covariance matrix is modified by another adaptive factor based on the residual vector. The performance of the suggested algorithm is analyzed in AV analysis and also with the conventional Kalman filter and a single transitive factor-based SHAFKF algorithm. It was testified that the SHAFKF algorithm was a suitable linear adaptive KF for minimizing drift and random noise of MEMS gyro signals in the static case.

To achieve agility and large-angle attitude maneuvers of spacecraft, Chapter 5 proposes a discrete-time nonlinear attitude tracking control in which the amplitude of the control input does not depend on the sampling period. The chapter considers discrete-time nonlinear attitude tracking control problems of spacecraft and derived a Euler approximation system with respect to tracking error; then a discrete-time nonlinear attitude tracking controller is introduced and the exact discrete-time system with a derived controller is analyzed. The effectiveness of the proposed control method is verified by numerical simulations.

Chapter 6 demonstrates that the MEMS gyroscope could be used as an effective tool for gait analysis. It could help to cut the cost of revealing underlying pathologies manifested by gait abnormalities. The chapter gives a close examination of human gait patterns (normal and abnormal) using gyroscope-based wearable technology, and experimental results show that foot-mounted gyroscopes could assess gait abnormalities in both temporal and spatial domains. Gait analysis systems based on MEMS gyroscopes are economical, convenient, and suitable to use in both the clinic and at home.

Xuye Zhuang
Department of Instrument Science and Technology,
School of Mechanical Engineering,
Shandong University of Technology,
China

Lianqun Zhou
CAS Key Laboratory of Biomedical Diagnostics,
Suzhou Institute of Biomedical Engineering and Technology,
Chinese Academy of Sciences,
China

Chapter 1

Introductory Chapter: Gyroscopes - Principles and Applications

Xuye Zhuang, Pinghua Li, Dongxing Li and Wentao Sui

1. Origins of gyroscopes

In ancient China, dating back to 500 BC, an interesting toy called "bamboo dragonfly" was invented to emulate the dragonfly [1]. The toy was very popular among children. When they rubbed this toy between their hands, it went flying into the air. Children had great fun experimenting to see whose bamboo dragonflies would fly the farthest and the highest. Bamboo dragonfly is an early type of gyroscope owning a history longer than 2500 years. But the name "gyroscope" did not appear until in the middle of 19th century. Gyroscope was created by a French physicist, Jean-Bernard-Léon Foucault [2]. He named his experimental instrument for Earth's rotation observation by joining two Greek roots: gyros meaning "circle or rotation" and skopeein meaning "to see." During his experiment to demonstrate the Earth's rotation, he found gyroscopes could maintain their original orientations in space regardless of Earth's rotation. This unique merit made gyroscopes are the perfect sensors to detect and measure the angular velocity of a rotational object, the deviation of a vehicle from its desired orientation. Since then gyroscopes were widely used in autonomous navigation systems. According to Encyclopedia Britannica, the first workable gyrocompass was developed by German inventor H. Anschütz-Kaempfe in 1908 [3]. It was invented to be used in a submersible. Then, in 1909, the first auto-pilot was created by an American inventor Elmer A. Sperry. It consisted gyroscopes which used to measure the rotation speed of the airplane. By collecting this information, gyroscopes could help stabilize the flight of the aircraft. In 1916, gyroscopes were used for assistant steering in a Danish passenger ship of a German company. Since then, gyroscopes became more and more popular in attitude control and navigation systems.

2. Development process of gyroscopes

The prototypes of gyroscopes designed by Léon Foucault were mechanical gyroscopes. The typical type of this kind of gyroscope was made by suspending a spinning relatively massive rotor inside three rings called gimbals. The basic principle of mechanical gyroscopes was the law of conservation of angular momentum: the tendency for the spin of a system to remain constant unless subjected to external torque. Mechanical gyroscopes are the most common or familiar type of gyroscope. Bamboo dragonfly fits into this category, which includes any gyroscope that relies on a ball bearing to spin. These types of gyroscopes are often used in navigation of aeronautic devices and vessels. However, due to the friction in the support bearings,

IntechOpen

imbalances inherent in the construction of the rotor, mechanical gyroscopes are typically noisier than other forms of gyroscopes, and their performances prone to drift with time, needing to be maintained frequently.

With the progress of science and technology, new kinds of gyroscopes based on new principles continuously came forth in our eyeshot and found their applications in our daily lives. Optical gyroscopes are the most successful ones, including ring laser gyroscopes and fiber-optic gyroscopes [4]. These devices send two beams of light around a circular path in opposite directions. If the path spins, a "fringe interference" pattern (alternate bands of light and dark) was detected that depended on the precise rate of rotation. They first appeared in the 1960s, following the invention of the laser and the development of fiber optics. Optical gyroscopes have the advantages of excellent measurement accuracy and having no moving parts and thus, no friction. The first ring laser gyroscope (RLG) was built in 1963 by Mecek and Davis. Owing to their high level of accuracy, cheap cost, high reliability, and easy maintenance, RLGs are perfect for integration in Inertial Navigation Systems. Today, RLGs have largely replaced their mechanical gyroscope predecessors in autopilot systems in aircrafts and guided missiles through missions where GPS is not safe to use. The global ring laser gyroscopes market is anticipated to reach a market value of US$ 948.3 million by 2026, growing at a compound annual growth rate of 3.5% during the forecast period 2018–2026 [5].

Fiber-optic gyroscope (FOG) is another successful optical gyroscope [4]. It was first proposed and studied in 1970s and was initially considered to be devoted to medium-level applications. In 1978, McDonald Company developed the first practical FOGs, and in 1980, Bergh et al. devised the first all-fiber optic gyroscope. Since then, FOGs have experienced a period of rapid development, the angular velocity measurement accuracy has been improved to 0.001°/h, reaching the strategic level of performance and surpassing the ring laser gyroscope in terms of deviation noise and long-term stability. In addition to no moving and wearing parts, FOGs also have the advantages of small size, light weight, large dynamic range, and flexible design, which mean the performance of a fiber-optic gyroscope can be adjusted by altering the length and diameter of its coil. FOGs are now gradually evolving in the direction of low cost, high reliability, and long service life. FOGs are broadly used in inertial navigation systems in guided missiles, aircrafts, and vehicles for attitude measurement and navigation. The global market of FOGs is expected to worth US$ 948.3 million by 2022, growing at a compound annual growth rate of 3.61% during the period 2016–2022 [6].

MEMS gyroscope is the most successful commercial modal for angular velocity sensors [7]. MEMS gyroscopes are based on micro electrical mechanical systems (MEMS) technology and are very suitable to mass production. Benefiting from the advancements in MEMS technology, MEMS gyroscopes own their most prominent advantages: cheap, small, and light. The common type of MEMS gyroscope is made of silicon, with a massive object suspended in the air by an anchor or some springs. When a gyroscope is in operation, the suspended structure keeps vibrating. As the gyroscope experiences a rotation relative to its reference, a force called Coriolis force will act on the suspended structure and causes it to move in a direction perpendicular to its vibrating direction. This movement is proportional to the rotation speed and converts into electrical signals that can be amplified and read by a microcontroller, the angular velocity is then ascertained. Once the accuracy of MEMS gyroscopes was much lower than that of their competitors, such as optical gyroscopes and mechanical gyroscopes. So, their applications were restricted to low-end. In the last 10–15 years, the precision of MEMS gyroscope improves drastically and achieves the tactical grade level (0.1°/h). On the benefits of the small size, low costs, and light weight and due to its improved precision and environmental stability, MEMS gyro is applied in more and more areas, such as consumer electronic devices, wearable devices, automotive safety, personal and vehicle navigation systems, robots controlling, stability controlling system, video games, and toys. MEMS gyroscope is also entering markets that were previously dominated by optical gyroscopes, now they have been the successful mass-produced commercial productions. The MEMS gyroscope market is growing rapidly and is expected to witness a CAGR of 9.48% during the period between 2020 and 2025 [8].

Gyroscope is now showing an encouraging momentum of accelerated development. New gyroscopes based on new principles, new fabrication process, and new materials are spring up like mushrooms, such as microscale hemispherical resonator gyroscopes, surface acoustic wave (SAW) gyroscopes, electrostatic (suspended) gyroscopes, integrated optic gyroscopes, superfluid gyroscopes, diamond gyroscopes, atomic gyroscopes, etc. [9–12]. The application field of gyroscope is also expanding rapidly, and it has covered almost every aspect of human lives, from aircrafts to vehicles, to wearable medical devices, to our smartphones, and to children's toys. Gyroscopes have been the most powerful tools of human beings to explore our living environment (ocean exploration, space exploration, planetary exploration, and even exoplanet exploration in the future) to improve the quality of our daily lives. Gyroscope is old, but also is young. With the efforts of researchers and scientists, gyroscopes are becoming more and more prosperous and benefit all humankind.

Acknowledgements

This work was/supported by the Taishan Scholars Program of Shandong Province No. tsqn201909108.

Author details

Xuye Zhuang*, Pinghua Li, Dongxing Li and Wentao Sui
School of Mechanical Engineering, Shandong University of Technology, Zibo, China

*Address all correspondence to: zxye@sdut.edu.cn

References

[1] Toys Invented in Ancient China. Classroom. Available from: https://classroom.synonym.com/toys-invented-ancient-china-10930.html [Accessed: 18 March 2020]

[2] Gyroscope. Available from: http://physics.kenyon.edu/EarlyApparatus/Mechanics/Gyroscope/Gyroscope.html [Accessed: 18 March 2020]

[3] Gyroscopes have a huge variety of uses, find out how they work in our latest Tech Explained. Tech Explained: Gyroscopes. Available from: https://www.springwise.com/tech-explained-gyroscopes/ [Accessed: 20 November 2018]

[4] Shamir A. An Overview of Optical Gyroscopes Theory, Practical Aspects, Applications and Future Trends: Technical paper: Adi Shammir; 2006

[5] Ring Laser Gyroscope Market-Snapshot. Transparency Market Research. available from: https://www.transparencymarketresearch.com/ring-laser-gyroscope-market.html [Accessed: 15 March 2020]

[6] Fiber Optics Gyroscope Market. Markets and Markets. Available from: https://www.marketsandmarkets.com/Market-Reports/fiber-optic-gyroscope-market-99825571.html [Accessed: 14 March 2020]

[7] Trusov AA. Overview of MEMS Gyroscopes: History, Principles of Operations, Types of Measurements. Irvine, USA: University of California; 2011

[8] MEMS Gyroscope Market - Growth, Trends and Forecast (2020-2025). Mordor Telligence. Available from: https://www.mordorintelligence.com/industry-reports/mems-gyroscope-market [Accessed: 15 March 2020]

[9] Fang J, Qin J. Advances in atomic gyroscopes: A view from inertial navigation applications. Sensors. 2012;**12**(5):6331-6346

[10] Zhuang X, Chen B, Wang X, et al. Microscale polysilicon hemispherical shell resonating gyroscopes with integrated three-dimensional curved electrodes. Journal of Physics: Conference Series. 2018;**986**:012022

[11] Oh H, Lee KJ, Lee K, et al. Gyroscopes based on surface acoustic waves. Micro and Nano Systems Letters. 2015;**3**(1):1

[12] Armenise MN. Emerging gyroscope technologies. In: Advances in Gyroscope Technologies. Berlin, Heidelberg: Springer; 2010. pp. 103-108

Chapter 2

IFOG and IORG Gyros: A Study of Comparative Performance

Ramón José Pérez Menéndez

Abstract

In this revision work, firstly classical structure and main performance parameters of interferometric fiber-optic gyroscope (IFOG) and integrated optics passive resonator gyroscope (IORG) are reviewed. Then, the main advanced models and performance parameters of these two types of rotation-rate inertial sensors are described, and finally the design trends of both types are analyzed. Taking as reference the performance parameters analyzed above, a comparative analysis between manufactured IFOG and IORG units of close geometrical dimensions is realized. This analysis leads ranking these devices into six classical levels of inertial performance: strategic grade, navigation grade, high-end tactical grade, tactical grade, industrial low-end tactical grade, and consumer grade. This classification allows to deduce the main application areas of both kinds of devices. This way, the impact of these sensors in applications such as aeronautics, aerospace navigation, mechanical micro-fabrication, tactical weapons, or, more recently, robotics can be disclosed.

Keywords: interferometric fiber-optic gyroscope (IFOG), integrated optics passive resonator gyroscope (IORG), optical passive ring resonator interferometer, single-mode fiber (SMF), silicon wire waveguide

1. Introduction

Applications as guidance, navigation, and control systems in aircrafts, spacecrafts, and attitude systems in terrestrial vehicles, to give some examples, require compact, low cost, and reliable inertial navigation systems (INSs), which are equipped with increasingly accurate gyroscopes. For this reason, gyroscopes (in what follows from here, gyros) are key elements, which are essential to obtain the desired sensitivity for all above applications. Gyros having a dynamic range up to $\pm 1500°/s$ are required in both space and terrestrial vehicle navigation inertial measurement units (IMUs). Attitude and heading systems in aircraft and precision-spacecraft INSs require a gyro resolution typically on the order of 1°/h and 0.01–0.001°/h, respectively. Optical gyros based on Sagnac effect are the key components of IMUs which are widely used in the above mentioned applications [1, 2]. Currently, the most widely used gyro technology for high-performance gyro systems is the optical fiber-based technology, specifically the technology based on interferometric fiber-optic gyro (IFOG). However, the counterpart of the IFOG made entirely in silicon wire waveguides on silicon-on-insulator (SOI) platform, called photonic integrated circuits (PICs) and, specifically, the integrated optics passive resonator gyro (IORG), has many advantages such as high robustness, theoretical sensitivity, and superior reliability due to its inherent characteristics of miniaturized structure, all-solid-state, and the combination between integrated optics and well-known CMOS fabrication technology [3, 4]. Thus, IORG has been considered as the next generation of resonant micro-optical gyros (RMOG) and a promising candidate in the field of inertial navigation [3, 5]. In particular, IORGs are very promising in terms of performance parameters such as low cost, compactness, light weight, and high reliability.

Both of them (IFOGs and IORGs) are based on Sagnac effect, which generates a phase or frequency difference proportional to the angular rate when two counter-propagating light beams in an optical resonant cavity suffer a rotation. Sagnac effect has also been demonstrated in semiconductor ring laser gyros (SRLGs) [1, 6]. Passive optical resonators with the laser source external to the ring resonator are particularly attractive because they show high performance and overcome some issues of active devices, mainly lock-in effect and mode competition. So far, best demonstrated resolution for the IFOG is 0.0002°/h [IXSea, FOG Marins] for a strategic-grade unit with 5 km length and 200 mm fiber coil diameter [7], while for the IORG, the best value achieved is 1°/h [IntelliSense Corp., VIGOR] for a temperature-compensated high-end tactical-grade unit with weight < 100 g and volume < 5 cc. made on a silicon wire waveguide ring resonator [8].

This work is structured as follows: Sections 2 and 3 deal with the main structure and configurations of IFOG and IORG, respectively. Section 4 examines the performance-grade classification and parameters of IFOG and IORG. Section 5 collects the main performance parameters of IFOG and IORG units made or designed by the most important world manufacturers and laboratories, respectively, enabling a comparative study between them. Section 6 presents the main design advances, trends, and optimization issues in the IFOG and IORG cutting-edge engineering. Finally, Section 7 is devoted to extract the main conclusions of this work.

2. IFOG: basic structure and configurations

The interferometric fiber-optic gyro is to date a mature technology and was originally designed as a low-cost alternative to the ring laser gyro (RLG). Surprisingly, today IFOG can substitute the RLG both in terms of manufacturing costs and that of performance, gaining prominence in a series of military and commercial applications [9–12]. The studies provide that the developments in solid-state optics and fiber-optic technology could lead up to 0.0001°/h ultimate value in resolution performance for IFOG units even for small-size designs. The IFOG is based on the Sagnac effect within an open optical path realized by a N-turn fiber-optic coil when two independent counter-propagating light modes are externally introduced from a broadband laser source through its two ends, respectively (see **Figure 1**). This causes an interference pattern between the CW and CCW light beams to be collected in a photodetector with a phase shift given by the following equation:

$$\phi_S = \frac{2\pi L D}{\lambda_0 \, c_0} \Omega \tag{1}$$

where L and D are length and diameter of fiber-optic sensing coil, respectively; λ_0 and c_0 are wavelength and speed of light source in vacuum, respectively; and Ω is the rotation rate.

DOI: http://dx.doi.org/10.5772/intechopen.89957

For a fixed length L by fixing the coil diameter D, the sensitivity of the sensor can be improved by increasing the total coil length L by adding a high number of turns N taking into account an upper limit due to the fiber attenuation. Also working at 1310 nm wavelength instead of 1550 nm could help to improve the sensitivity of the sensor. From **Figure 1** it can be clearly seen that IFOG has a passive configuration because the laser source is located externally to the sensing coil. In this system, the two counter-propagating light beams travel through the core of a single-mode optical fiber (SMF) under the total internal reflection phenomenon. As the core diameter of such an optical fiber is only about 8 μm, the spot size of the interference signal can only be coupled to a small area at the end of the fiber loop, for example, on the small detection area of a photodetector. So that, this interference signal affects only one or two interference fringes whose intensity can be evaluated by the following expression:

$$I(\phi) = I_0(1 + \cos\phi) \tag{2}$$

where I_0 is the amplitude of each of the two counter-propagating beams and ϕ is the optical phase difference between them. **Figure 2** represents the variation of light

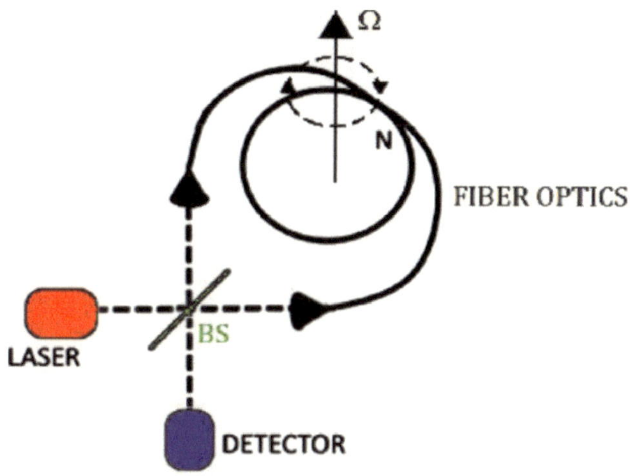

Figure 1.
Basic structure of the IFOG.

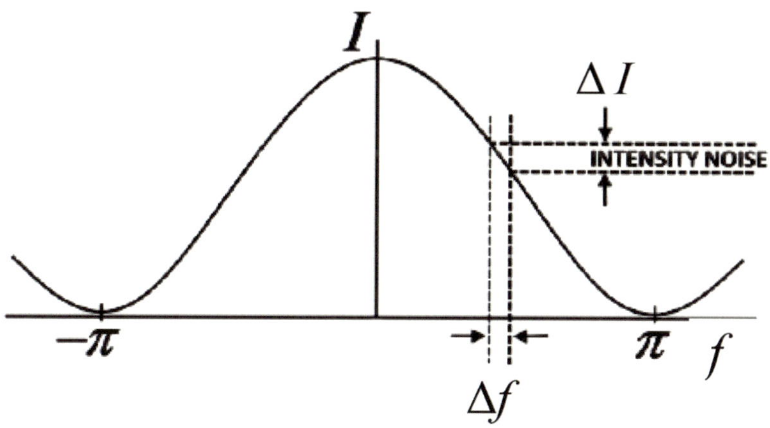

Figure 2.
Two-beam interference response curve as a function of ϕ (phase difference).

intensity along a single interference fringe as a function of ϕ. Notice the output intensity noise produced when the phase difference is detected with a phase error $\Delta\phi$. Phase noise sources and their influence are treated in Ref. [13].

Figure 3 shows the minimal open-loop IFOG unit configuration. This configuration includes apart fiber sensing coil, a superluminescent (SLD) broadband light source, two (2 input × 2 output) directional couplers, one linear polarizer, one photodetector, and one electro-optic phase modulator made of piezo ceramic tube (PZT) or lithium-niobate electrodes. The function of electro-optical phase modulator is to provide a controlled phase shift which adds to the Sagnac phase shift produced by the rotation onto the system.

This way, the signal detected by photodetector can be demodulated with some ease to recover by electronic means the Ω rotation-rate value which affects the whole system. A reduced minimal configuration fiber-optic gyro for land navigation applications can be found in Ref. [14]. Then two main options can be adopted for IFOG: open-loop configuration and closed-loop configuration. **Figure 4** represents one typical bulk optics open-loop IFOG configuration, the dotted block being usually made on an integrated optical chip (IOC). In this configuration, the rotation-rate information is recovered by the electrical output signal of photodetector after a previous demodulation process.

This configuration guarantees the reciprocity of the system. This implies that the two counter-propagating beams have exactly equal amplitude and phase at output when no rotation affects the system. However an error phase shift can be present on interferometric signal collected by the photodetector. So to reduce phase difference error and increase the resolution of the sensor, it is necessary to reach the

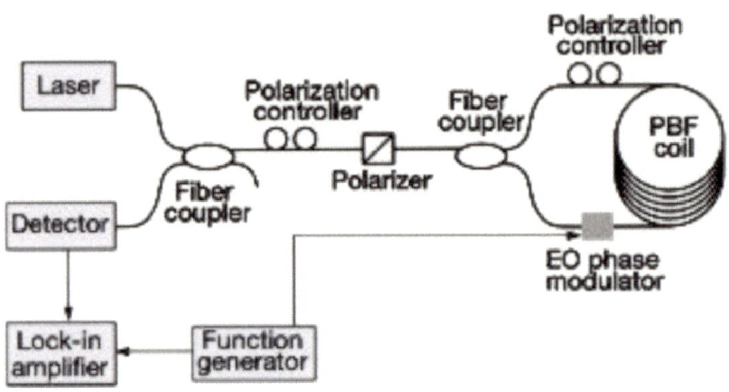

Figure 3.
Minimal open-loop IFOG unit configuration.

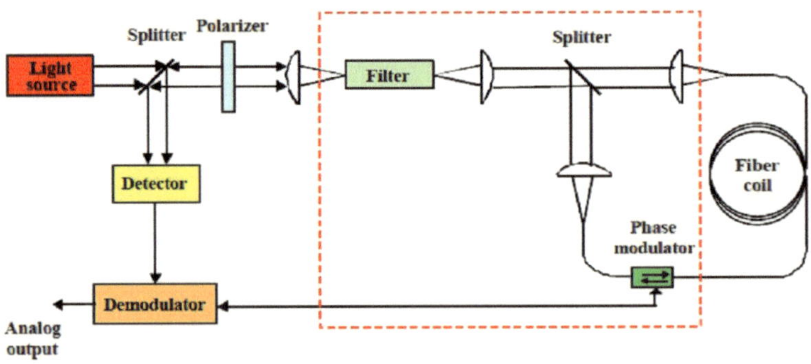

Figure 4.
Typical bulk optics IFOG open-loop configuration.

reciprocity condition as exactly as possible. Optical light source used is a broadband laser source to reduce the noise signal due to Rayleigh scattering within the fiber. As fiber sensing coil, either standard or rare-earth doped optical fiber can be used, but polarization-maintaining fiber is needed in both cases to ensure that just one polarization mode exists within it. The optical round trip experienced by the two counter-propagating beams is as follows. The light beam from source is collimated and coupled into an optical straight path and, then, passes through an optical system composed of one beam splitter, one linear polarizer, and a filter between two convergent lenses to select only one propagated mode within the fiber coil. The second beam splitter is used to split the original beam from the source and create two counter-propagated CW and CCW beams into the fiber coil. Then, after having traveled through the fiber loop, the CW and CCW beams recombine into the interferometer after passing again through the polarizer and reflecting onto the second face of splitter. Then, the photodetector collects the produced interferometric signal. The phase modulator is used to apply a sinusoidal or square-wave dynamic phase bias to light path, thus increasing the sensitivity of the sensor [15]. When modulation frequency of phase bias is high enough, electronic noise is avoided. Finally, an electronic demodulation circuit is needed to extract the magnitude and sign of rotation rate. Main advantages for the open-loop IFOG scheme are a few number of optical and electronic components and, hence, low price, good sensitivity, long lifetime, high reliability, and low power consumption. The main disadvantages of this configuration are high fiber coil length (100–3000 m) to increase sensitivity, electronic drift of analogical components, and disturbing influence of temperature and environmental conditions.

A more advanced design is achieved by closing the measurement loop by means of a feedback signal becoming into the so-called IFOG closed-loop configuration. The general scheme of a closed-loop IFOG is depicted in **Figure 5**. Very high performance was obtained for closed-loop IFOG configuration with respect the open-loop one [16]. In this scheme, the output signal of demodulator circuit passes through a servo amplifier which drives a phase transducer placed in the interferometer path. The total phase shift becomes equal to zero because the phase transducer introduces a nonreciprocal phase shift that is equal, by in the opposite sign, to that generated by Sagnac phase shift induced by rotation. The output of the system is then the output of the phase transducer. The main advantage of this configuration is the insensitivity to the laser source amplitude variations and the electronic circuitry gain because the system is always operated at zero total phase shift. This

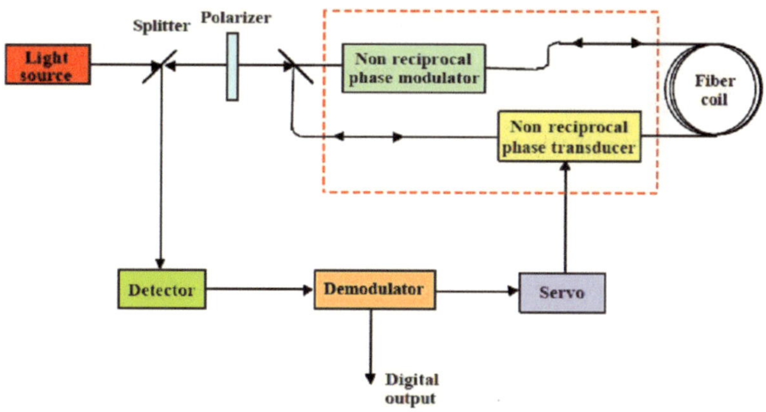

Figure 5.
Typical closed-loop IFOG configuration.

brings very small drift, from 0.001°/h up to 0.01°/h On the other hand, the output scale factor linearity and stability depends only on the phase transducer accuracy.

Eq. (3) allows the calculation of photon-shot-noise photocurrent at photodetector (I_{sn}), which is the minimum photodetector output current corresponding to a given level of optical input power:

$$I_{sn} = \sqrt{\frac{e^2 \eta \lambda}{hc} P_{\max} \Delta f} \tag{3}$$

Here, $P_{\max}$ is the maximum optical power incident at photodetector, Δf is the minimum bandwidth detectable by the photodetector, and e, η, λ, h, and c are the electron charge, the quantum efficiency of photodetector, the operating vacuum wavelength, the Planck constant, and the vacuum speed of light, respectively. This way, in accordance with photon-shot-noise photodetector current, the $\delta\Omega$ threshold sensitivity of gyro sensor (that is to say, the minimum rotation rate which this gyro is able to measure) can be calculated by Eq. (4):

$$\delta\Omega = \left(\frac{hc^2}{\pi e \eta L D P_{\max}}\right) I_{sn} \tag{4}$$

where L and D are the length and diameter of sensing fiber coil, respectively, and the other parameters are the same as in Eq. (3). From this expression it can be clearly seen that the greater the product L × D, the better the sensitivity of the detector. Thus, typical performance reported for open-loop and closed-loop IFOG configurations are ±100°/hr [17–21]. dynamic range, bias drift between 0.001°/h and 0.2°/h with drift stability between 0.0005°/h and 0.01°/h, angle random walk (ARW) between $0.004°/\sqrt{\text{hr}}$ and $0.04°/\sqrt{\text{hr}}$, and bandwidth from 20 to 100 Hz. Those IFOGs have long been used for land navigation applications mainly due to their extreme robustness, and also they are also commercially available for use in space applications due to their high reliability and cost-effectiveness.

3. IORG: fundamentals and main configurations

The usage of RFOGs instead of IFOGs is the first step that allows to reduce the fiber-optic coil length, thus leading to lower dimensions. Then, in the era of miniaturization, the possibility of integrating optical waveguides different than optical fibers leads to even smaller geometrical dimensions. Thus, RMOG is a promising candidate for applications requiring small, light, and robust gyros. The first design of an RFOG was made by S. Ezekiel and S.R. Balsamo at M.I.T. in 1977. In this design, the difference Δf between CW and CCW frequency resonances of the cavity is given by the following equation [22, 23]:

$$\Delta f = \left(\frac{4A}{\lambda P}\right) \Omega \tag{5}$$

where A is the area enclosed by the cavity, P is the perimeter of the cavity, λ is the vacuum wavelength, and Ω is the rotation rate affecting the system. The precision with which Δf can be measured depends on the Q factor of the cavity. This way, the minimum rotation rate that this gyro is able to measure can be calculated by Eq. (6):

$$\delta\Omega = \frac{1}{QD\sqrt{P_{pd}}}\sqrt{\frac{2hc^3}{\lambda\eta\tau}} \tag{6}$$

here Q is the Q factor of the cavity, D is its diameter, P_{pd} is the input optical power incident on photodetector, τ is the integration time, and h, c, λ, and η are the same as in Eq. (4). A new step to achieve the miniaturization of the gyro is to perform the passive ring resonator by means of an integrated optical waveguide made of high-index-contrast materials like silica-on-silicon, silicon-on-insulator, III–V semiconductors (InP), or silicon nitride (Si_3N_4). Then, this solution gets to what is called the IORG. Thus, an IORG can be formed by a ring resonator that includes an optical waveguide having a ring shape and one or two straight bus waveguides (see **Figure 6**). The bus and the ring waveguides are coupled by the evanescent field. When the ring is used for rotation sensing, it is necessary to launch two input signals (CW, CCW) simultaneously in the bus waveguides to excite the ring resonator cavity for both the CW and CCW propagation directions for the Sagnac effect to be applied.

If a two-bus waveguide approach is used, the two input beams can be launched in two different bus waveguides or in the two opposite ends of the same bus waveguide. Consequently there are two possible configurations for the excitation of the cavity and resonance frequency measurement. In the first case, output ports are called through ports (**Figure 6(a)** and **(c)**), whereas in the second one, output ports are the drop ports (**Figure 6(b)**). Using a one-bus waveguide architecture, each end of the bus can be utilized either as input or output port. In this case, two circulators or switches have to be used at both ends of the bus to excite the resonator in CW and CCW directions and to monitor the spectral response at the respective through port (**Figure 6(c)**). To minimize the bias drift of the gyro, the two beams coupled to the resonator must have the same optical power amplitude or as similar as possible with very reduced tolerance.

The conventional configuration of an IORG includes a narrow linewidth laser source, a high Q factor ring resonator, an optoelectronic processing unit, two photodetectors, and an electronic readout unit (**Figure 7**). The sensor can be manufactured by using hybrid or monolithic photonic integration. Hybrid integration has as main problems the optical alignment of all components and the high value of optical power losses. Monolithic integration has the advantages of the absence of optical alignment issues, the higher robustness and compactness, the minor dimensionality, and the lower optical power consumption.

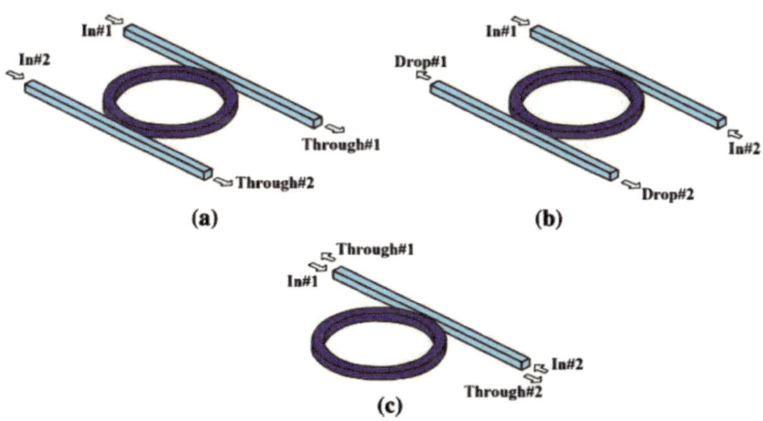

Figure 6.
Integrated ring resonator coupled with one (c) or two (a), (b) waveguides (sensing element of IORG).

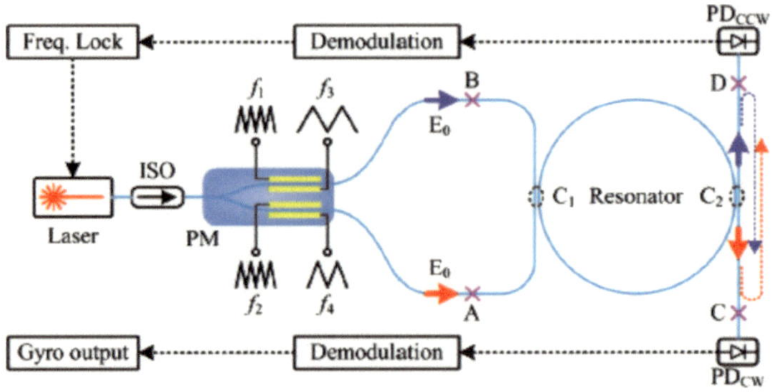

Figure 7.
Conventional configuration of an IORG, including a narrow linewidth laser source, an optical isolator, optoelectronic components for demodulation processing, a waveguide ring resonator, two detectors, and electronics readout unit.

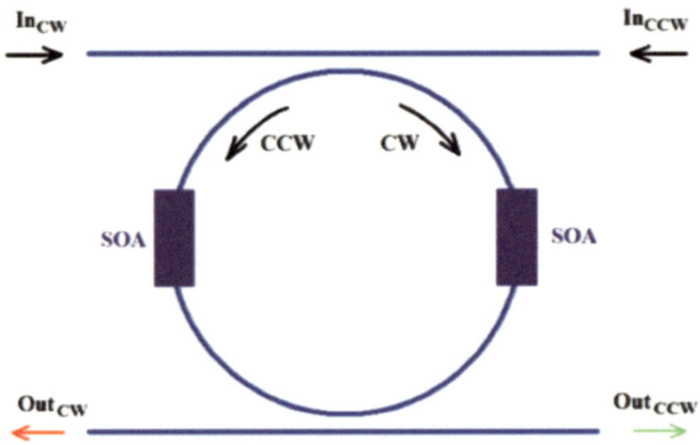

Figure 8.
Two SOAs are incorporated within ring resonator to compensate propagation loss.

The dimensions of the integrated passive resonator influence gyro scale factor, and, in time, the cavity Q factor depends on loss and resonator length. This way, to achieve $\delta\Omega < 5°/\text{h}$ and ARW $< 0.02°/\sqrt{\text{h}}$, a Q factor around 10^6 and a resonator length in the range of centimeters are at least required [24].

Maximum achieved Q factor in SOI ring resonators is around 1.5×10^5. This limitation in Q factor makes very difficult to realize a passive integrated optical gyroscope having $\delta\Omega < 5°/\text{h}$ by using a SOI ring resonator. On the other hand, silica-on-silicon technology allows very low loss (<0.1 dB/cm) operating at 1.55 μm, so that waveguides made on this technology are more suitable for IORG engineering. Propagation losses around 0.02–0.03 dB/cm have been achieved for low-index contrast silica-on-silicon waveguides ($\Delta < 1\%$). As bending loss suffered by these waveguides exponentially decreases with curvature radius, to achieve negligible bending loss, a curvature radius larger than a few millimeters is required. Some ring resonators employing a 5 μm × 5 μm squared core having a very large quality Q factor (2.4×10^7) and operating at 1.55 μm wavelength have been fabricated in silica-on-silicon technology. To further enhance the Q factor of resonators in silica-on-silicon technology, the hybrid integration of two semiconductor optical amplifiers (SOAs) within a silica-on-silicon ring resonator was proposed (**Figure 8**).

For a ring radius of 10 mm and a cross-coupling coefficient of 0.001 between, a Q factor as high as 2.9×10^8 was calculated, neglecting the effect of spontaneous emission noise. Then, with this ring resonator configuration, an IORG unit exhibiting 10°/h bias drift was achieved. A spiral resonator having a total length of 42 cm and a footprint of 20 cm^2 was designed in 2012 by Ciminelli et al. with an estimated bias drift equal to 0.2°/h, which is the best value reached to date for this kind of IORG [25]. A new trend of design over InP waveguide technology is emerging to improve the scale of integration of IORG, to come to make a true gyro-on-a-chip (GoC). However, substantial improvements have to be found in this technology because the best value calculated so far reaches a resolution of 10°/h.

4. IFOG and IORG: performance grade and parameters

Table 1 collects the six levels of gyro performance-grade classification together with its characteristic rotation-rate resolution. As it can be seen, each performance level includes scales by two orders of magnitude. For each of them, the main areas of application are detailed on the right column.

What follows next is a concise description of each gyro performance parameter. Four main gyro performance parameters will be considered here, namely, bias stability, scale factor linearity, angle random walk, and dynamic range. For one more deep analysis and exhaustive information about the definition of gyro performance parameters, the reader should consult [26].

4.1 Bias stability

Bias instability is the measurement of bias offset at any constant temperature and ideal environment. It can be measured using the Allan variance technique. Bias instability introduces errors that may not be easy to calibrate. Its influence is greater on longer measurement periods, so that bias instability is one of the most critical parameters in the gyro selection process for applications that require excellent accuracy over long time. Therefore, two values of bias stability are usually considered: (1) the long-term bias drift for long integration time values, say 1 min or 1 h, and (2) the short-term bias drift, say 1 s.

4.2 Scale factor linearity

The linearity of the scale factor is the maximum separation with respect to the linear variation of the rotation rate expressed in ppm (parts-per-million) or % (parts-per-cent).

Performance grade/bias stability range		**Applications**
Consumer	30–1000°/h	Motion interface
Industrial and low-end tactical	1–30°/h	Ammunitions and rocket guidance
Tactical	0.1–30°/h	Platform stabilization
High-end tactical	0.1–1°/h	Missile navigation
Navigation	0.01–0.1°/h	Aeronautics navigation
Strategic	0.0001–0.01°/h	Submarine navigation

Table 1.
IFOG and IORG: performance-grade classification and respective applications.

4.3 Angle random walk (ARW)

In the output of a gyro, there is always a broadband white noise element. Angle random walk describes the error resulting from this noise element and can be evaluated using the Allan variance technique. Active elements of the gyro are the major contributors to random noise (laser diode and photodiode for optical gyroscopes and the vibrating beam and detection electronics for MEMS). Noise is one of the most important differences between optical and MEMS gyro performance, resulting in different precision and accuracy in measurements.

4.4 Dynamic range

Dynamic range is the maximum excursion of rotation rate that the gyro is able to measure with a maximum error specified by the rotation-rate drift (bias stability).

5. IFOG and IORG performance parameters: a comparative analysis

Next, in **Table 2**, the performance parameters of 33 IFOG, RFOG, RLG, or IORG gyros are collected for comparison purposes. Five gyro performance parameters are considered here: bias stability, scale factor linearity, ARW, dynamic range, and dimensions/weight/response time. For each type of gyro unit, either manufactured or designed, the manufacturer and the main areas of application are also specified. In the case of the units manufactured, the performance parameters were obtained from the technical manufacturer data sheets, while for the units designed and tested in the laboratory, the data have been obtained from the referenced scientific publication.

6. IFOG and IORG: advanced design, trends, and optimization

In recent years, several groups of researchers throughout the world are devoting great effort in the development of high-performance resonant micro-optical gyros (RMOG). All of these RMOG designs are based on a waveguide/ring resonator structure acting as the sensor element of rotation rate. The main variants of this design are focused either on ultrahigh Q silica-on-silicon ($Q = 1.5 \times 10^6$) or InP ($Q \geq 10^6$) ring resonators or on Si waveguide/photonic crystal (PhC) ring resonator ($Q = 7 \times 10^8$). More explicitly, the very promising research field of IORGs aiming at the development of optoelectronic ultracompact and high-performance gyros compliant with the requirements of aerospace and defense industry is recently focused onto five technological approaches that are being explored: (1) SRLGs, (2) RMOGs based on ultrahigh Q silica resonators, (3) InP gyro-on-a-chip, (4) the gyro configuration based on the ring cavity with a Bragg grating in the resonant path, and (5) the multi-ring-cavity gyro. For all of them, main efforts focus on design improvement, efficient modulation technique, and resolution enhancement. It is expected that a gyroscope on-a-chip prototype will be developed soon. If the characterization of that prototype will be successful, it is expected that the gyro-on-a-chip will have a very notably impact on the aerospace and defense navigation applications.

IFOG/IORG gyro technology	Performance grade/bias stability range		IFOG/IORG parameters					Manufacturer [model]/researcher ref. [X]	Applications
			Bias stability	Scale factor linearity	Dimensions/weight/response time	ARW	Dynamic range		
IFOG	High-end tactical	0.1–1°/h	≤1°/h	≤10,000 ppm	18.8mm × 19.1 mm × 24.6 mm (L × W × H)/25 g	$\leq 0.05^{\circ}/\sqrt{h}$	±200°/s	Northrop Grumman LITEF [G-2000]	Missile and platform stabilization
IFOG	Navigation	0.01–0.1°/h	≤0.07°/h	≤100 ppm	L = 200 m, D = 88.9 mm/750 g/no data	$\leq 0.056^{\circ}/\sqrt{h}$	±11.46°/s	Litton [LN-200]	Aeronautics navigation
IFOG	Navigation	0.01–0.1°/h	≤0.04°/h	≤100 ppm	210.8 × 119.9 × 79.8 (L × W × H)/1950 g	$\leq 0.002^{\circ}/\sqrt{h}$	±300°/s	Northrop Grumman LITEF [LR-240]	Image and platform stabilization
IFOG	Strategic	0.0001–0.01°/h	≤0.0009°/h	≤10 ppm	L = 1000 m, D = 76 mm/750 g	$\leq 0.0009^{\circ}/\sqrt{h}$	±250°/s	Litton guidance and control systems [7]	Space navigation
RLG	High-end tactical	0.1–1°/h	≤1°/h	≤1 ppm	Volume = 541 cm^3/900 g	$\leq 0.125^{\circ}/\sqrt{h}$	±358°/s up to ±1620°/s	Honeywell [HG1700IMU]	Missile navigation
IFOG	Strategic	0.0001–0.01°/h	≤0.0003°/h	≤1 ppm	L = 4000 m, D = no data/no data	$\leq 0.00010^{\circ}/\sqrt{h}$	No data	Honeywell [HPFOG]	Space navigation
IFOG	Strategic	0.0001–0.01°/h	≤0.0007°/h	≤15 ppm	L = 1500 m, D = 180 mm /no data	$\leq 0.00022^{\circ}/\sqrt{h}$	±30°/s	IXSea [FOG180]	Aeronautics navigation
IFOG	Strategic	0.0001–0.01°/h	≤0.0005°/h	≤30 ppm	L = 2000 m, D = 330 mm/no data	$\leq 0.00021^{\circ}/\sqrt{h}$	±15°/s	IXSpace [ASTRIX200]	Space navigation
IFOG	Strategic	0.0001–0.01°/h	≤0.0002°/h	≤15 ppm	L = 5000 m, D = 370 mm/no data	$\leq 0.00017^{\circ}/\sqrt{h}$	±15°/s	IXSea [FOG MarinsM3]	Submarine navigation
IFOG	Tactical	0.1–30°/h	≤5°/h	≤1000 ppm	L = 75 m, D = 58.42 mm/≤3 ms	$\leq 0.14^{\circ}/\sqrt{h}$	±204°/s	KVH [DSP1500digital]	Platform stabilization
IFOG	Tactical	0.1–30°/h	≤3°/h	<500 ppm	L = 125 m, D = 58.42 mm/no data	$\leq 0.10^{\circ}/\sqrt{h}$	±100°/s	KVH [DSP3000analog]	Platform stabilization

IFOG/ IORG gyro technology	Performance grade/bias stability range		IFOG/IORG parameters					Manufacturer [model]/ researcher ref. [X]	Applications
			Bias stability	Scale factor linearity	Dimensions/weight/response time	ARW	Dynamic range		
IFOG	Tactical	0.1–30°/h	≤1°/h	<500 ppm	L = 300 m, D = 58.42 mm/no data	≤0.067°/√h	±375°/s	KVH [DSP3000 digital]	Platform stabilization
IFOG	Navigation	0.01–0.1°/h	≤0.05°/h	≤500 ppm	L = 500 m, D = 45.97 mm/≤ 1.3 ms	≤0.013°/√h	±490°/s	KVH [DSP1750]	Aeronautics navigation
IFOG	Navigation	0.01–0.1°/h	0.10°/h (RMS)	50 ppm	L = 200 m, D = 81.2 mm/220 g	≤0.015°/√h	±1000°/s	Emcore [EMP-1]	Aeronautics navigation
IFOG	Strategic	0.0001–0.01°/h	0.005°/h (RMS)	25 ppm	L = 200 m, D = 83.8 mm/230 g	≤0.002°/√h	±1000°/s	Emcore [EMP-1.2 k]	Aeronautics/aviation
IFOG	Strategic	0.0001–.01°/h	0.0060°/h	≤300 ppm	L = 200 m, D = 78 mm/200 g	≤ 0.015°/√h	± 550°/s	Optolink [SRS-200]/[28]	Submarine navigation
IFOG	Strategic	0.0001–0.01°/h	0.0011°/h	≤200 ppm	L = 500 m, D = 100 mm/350 g	≤0.003°/√h	±250°/s ±1000°/s	Optolink [SRS-501]/[28]	Submarine navigation
IFOG	Strategic	0.0001–0.01°/h	0.0006°/h	≤100 ppm	L = 1000 m, D = 150 mm/900 g	≤0.0005°/√h	±550°/s	Optolink [SRS-1001]/[28]	Submarine navigation
IFOG	Strategic	0.0001–0.01°/h	0.00024°/h	≤30 ppm	L = 2000 m, D = 250 mm/ 1700 g	≤0.00026°/√h	±30°/s	Optolink [SRS-2000]/[28]	Space navigation
IFOG	Strategic	0.0001–0.01°/h	0.00008°/h	≤10 ppm	L = 5000 m, D = 250 mm/ 2500 g	<0.000069°/ √h	±12°/s ±550°/s	Optolink [SRS-5000]/[28]	Space navigation
IFOG	Tactical	0.1–30°/h	30°/h (RMS)	100 ppm	L = 75 m, D = 24 mm/30 g/ 20 ms	≤0.090°/√h	±300°/s	Fizoptika [VG091A]/[29]	Platform stabilization
IFOG	Tactical	0.1–30°/h	20°/h (RMS)	100 ppm	L = 100 m, D = 60 mm/45 g/ 10 ms	≤0.040°/√h	±300°/s	Fizoptika [VG949P]/[29]	Platform stabilization
IFOG	Tactical	0.1–30°/h	10°/h (RMS)	100 ppm	L = 125 m, D = 82 mm/120 g/< 20 ms	≤0.015°/√h	±250°/s	Fizoptika [VG095M]/ [29]	Platform stabilization

IFOG/IORG gyro technology	Performance grade/bias stability range		IFOG/IORG parameters					Manufacturer [model]/researcher ref. [X]	Applications
			Bias stability	Scale factor linearity	Dimensions/weight/response time	ARW	Dynamic range		
IFOG	High-end tactical	0.1–30°/h	1.00°/h (RMS)	100 ppm	L = 150 m, D = 129 mm/280 g	≤0.015°/$\sqrt{h}$	±60°/s	Fizoptika [VG035Q]/ [29]	Platform stabilization
IFOG	High-end tactical	0.1–1°/h	0.10°/h (RMS)	100 ppm	L = 200 m, D = 160 mm/320 g	≤0.007°/$\sqrt{h}$	±60°/s	Fizoptika [VG951]/ [29]	Missile navigation
RFOG	Low-end tactical	0.1–1°/h	0.10°/h (RMS)	No data	L = 100 m, D = 50.8 mm/no data	≤0.029°/$\sqrt{h}$	No data	Honeywell [RFOG], [30]	Commercial navigation
IORG	Low-end tactical	1–30°/h	10°/h	< 10,000 ppm	volume < 5 cm^3/weight < 100 g/no data	<0.1°/$\sqrt{h}$	±100°/s	IntelliSense [VIGOR], [8]	Ammunitions and rocket guidance
IORG	Low-end tactical	1–30°/h	1.432°/h	344.71 ppm	60 mm ring resonator Ø/no data/76 μs	0.8°/$\sqrt{h}$	±300°/s	Feng et al. [31]	Ammunitions and rocket guidance
IFOG	Consumer	30–1000°/h	≤180°/h	≤2000 ppm	No data/no data/50 Hz	≤0.04°/$\sqrt{h}$	±60°/s	Hitachi Cable [HOFG-1]	Robotics
IFOG	Low-end tactical	1–30°/h	≤1.00°/h	<2000 ppm	60 mm × 60 mm × 19.5 mm (L × W × H)/85 g	<0.1°/$\sqrt{h}$	±300°/s	Lockheed Martin (NEDAERO) [FOG-60, FOG-80]	AHRS aircraft, ground vehicles, robotics, platform stabilization (antennas)
IFOG	High-end tactical	0.1–1°/h	≤0.05°/h	<50 ppm	90 mm × 90 mm × 88 mm/ 655 g/440 Hz	<0.012°/$\sqrt{h}$	± 490°/s	Advanced Navigation [SPATIAL FOG]	Surveying applications, robot navigation, ground vehicle positioning
IFOG	Navigation	0.01–0.1°/h	< 0.02°/h (short-term) < 0.20°/h (long-term)	<150 ppm	Ø2.7" × 2" (11.5 in)/0.8 lb. (362.87 g)/>40 kHz	<0.0022°/$\sqrt{h}$	± 360°/s	Lockheed Martin (IFOS) [G5-G7-G8-G9 prototypes]	Aeronautics navigation

IFOG/ IORG gyro technology	Performance grade/bias stability range		IFOG/IORG parameters					Manufacturer [model]/ researcher ref. [X]	Applications
			Bias stability	Scale factor linearity	Dimensions/weight/response time	ARW	Dynamic range		
IFOG	Strategic	0.0001–0.01°/h	<0.0003°/h	<100 ppm	No data	<0.000053°/ $\sqrt{h}$	From 0.0015°/ h up to 1500°/ h	Lockheed Martin (Optiphase) [prototype's specification]	Submarine navigation, space positioning and navigation
IORG	Low-end tactical	1–30°/h	≤10°/h	< 10,000 ppm	Volume < 5 cm^3/ weight < 100 g/no data	<0.1°/$\sqrt{h}$	± 100°/s	IntelliSense [VIGOR], [8]	Ammunitions and rocket guidance
IORG	Low-end tactical	1–30°/h	1.432°/h	344.71 ppm	60 mm ring resonator Ø/no data/ 76 μs	0.8°/$\sqrt{h}$	± 300°/s	Feng et al. [31]	Ammunitions and rocket guidance
IORG	High-end tactical	0.1–1°/h	0.20°/h	No data	94.8 mm cavity length	0.00075°/$\sqrt{h}$	No data	Ciminelli et al. [24–32]	Aerospace/defense industry

Table 2.
IFOG and IORG gyro technology comparison in terms of performance parameters.

7. Conclusions

IFOGs have higher resolution performance than RFOG and IORG gyros. Therefore, IFOG technology is the best option for strategic-grade (0.0001°/hr), navigation-grade (0.001°/hr), or high-end tactical-grade (0.01°/hr) applications. Best RFOG designs reach high-end tactical-grade (0.01°/hr) or tactical-grade (0.1°/hr) performance, and they constitute a mature and tested technology for a large set of applications ranging from aircraft navigation up to platform stabilization. On the other hand, IORG technology is not yet mature, and over the last decade, it has experienced a vigorous development and refinement. Best results obtained experimentally in the laboratory for the performance of IORG prototypes are of 0.20°/h resolution and $0.00075°/\sqrt{hr}$ARW, respectively. As already mentioned above, several prototypes of RMOGs based on silica resonators have been already theoretically engineered, but the experimentally demonstrated performance is still at least one order of magnitude worse than that one demanded by aerospace and defense navigation applications. Therefore, an improvement of those kinds of gyros is needed to realize a significant impact on the market.

Acknowledgements

The author wishes to thank the UNED-Spain for technical support and assistance provided in the realization of this work.

Author details

Ramón José Pérez Menéndez
UNED-Spain, Lugo, Spain

*Address all correspondence to: ramonjose.perez@lugo.uned.es

References

[1] Armenise MN, Passaro VMN, De Leonardis F, Armenise M. Modeling and design of a novel miniaturized integrated optical sensor for gyroscope applications. Journal of Lightwave Technology. 2001;**19**(10):1476-1494

[2] Ciminelli C, Peluso F, Armenise MN. A new integrated optical angular velocity sensor. In: SPIE Conference on "Integrated Optics: Devices, Materials and Technologies IX", San Jose (USA), January. 2005

[3] Suzuki K, Takiguchi K, Hotate K. Monolithically integrated resonator microoptic gyro on silica planar lightwave circuit. Journal of Lightwave Technology. 2000;**18**(1):66-72

[4] Li G, Winick KA, Vikjaer EAJ. Design, fabrication and characterization of an integrated optic passive resonator for optical gyroscopes. In: ION 60th Annual Meeting 2004. June 2004

[5] Ford C, Ramberg R, Johnson K. Cavity element for resonant micro optical gyroscope. IEEE AES Systems Magazine. December 2000;**15**(12):33-36

[6] Armenise MN, Armenise M, Passaro VMN, De Leonardis F. Integrated optical angular velocity sensor. In: Politecnico di Bari, European Patent EP 1219926. 2000

[7] Korkishko Y, Fedorov V, Prilutskiy VE, Ponomarev VG, Morev IV, Obuhovich DV, et al. Investigation and identification of noise sources of high precision fiber optic gyroscopes. In: 20th Saint Petersburg International Conference on Integrated Navigation Systems. 2013

[8] Monovoukas C, Swiecki A, Maseeh F. Integrated optical gyroscopes offering low cost, small size and vibration immunity. In: Proceedings of SPIE 3936, Integrated Optics Devices IV (24 March). 2000

[9] Pavlath G. Fiber optic gyros: The vision realized. In: 18th International Conference on Optical Fiber Sensors, Cancun, Mexico. 2006

[10] KVH Industries Inc. An update on KVH fiber optic gyros and their benefits relative to other gyro technologies. March 2007

[11] Divakaruni S, Sanders S. Fiber optic gyros—A compelling choice for high accuracy applications. In: 18th International Conference on Optical Fiber Sensors, Cancun, Mexico. 2006

[12] Gaiffe T. From R&D brassboards to navigation grade FOG-based INS: The experience of photonetics/IX sea. In: 15th Optical Fiber Sensors Conference Technical Digest. Vol. 1. May 2002. pp. 1-4

[13] Wang W, Wang J. Study of modulation phase drift in an interferometric fiber optic gyroscope. Optical Engineering. 2010;**49**(11): 114401

[14] Emge S, Bennett S, Dyott R, Brunner J, Allen D. Reduced minimum configuration fiber optic gyro for land navigation applications. IEEE Aerospace and Electronic Systems Magazine. April 1997;**12**(4):18-21

[15] Komachia M, Sonobe H, Oho S, Ohbu K, Yuhara T, Hizuka H. Secondary-phase modulation method for open loop fiber optic gyroscopes. Applied Optics. 1996;**35**(9):3719-3725

[16] Lefèvre HC, Marten P, Morrise J, Simonpieti P, Vivenot P, Arditty HJ. High dynamic range fiber gyro with all digital signal processing. In: DePaula RP, Udd E, editors. Fiber Optic and Laser

Sensors VIII, Proceedings - Society of Photo-Optical Instrumentation Engineers, 1367. 1990. pp. 72-80

[17] Pavlath GA. Fiber optic gyro based inertial navigation systems at Northrop Grumman. In: Optical Fiber Sensors Conf. Tech. Dig., OFS 2002, Vol. 1. 2002. p. 9

[18] Sanders SJ, Strandjord LK, Mead D. Fiber optic gyro technology trends—A honeywell perspective. In: Optical Fiber Sensors Conf. Tech. Dig., OFS 2002, Vol. 1. 2002. p. 9

[19] Dyott RB, Bennett SM, Allen D, Brunner J. Development and commercialization of open loop fiber gyros at KVH industries. In: IEEE Optical Fibers Sensors Conference, 15th OFS 2002. 2002. pp. 19-22

[20] Jilmore JP, Freier L, Nolan E, Perlmutter M, Bowser M, Maglieri J. Three-axis nested fiber optic gyroscope. Available at: http://www.fibersense.com [Accessed: 21 April 2019]

[21] Dollon M, Cros G, Sevellec A, Antoine P, Muller G, Willemenot E, et al. A new family of IMU based on IFOG technology. In: Proceedings of the V ESA on Spacecraft—Guidance, Navigation and Control, SP 516, Frascati, Italy, 22–25 October. 2002. pp. 41-45

[22] Ezekiel S, Balsamo SR. Passive ring resonator gyroscope. Applied Physics Letters. 1977;**30**(9):478-480

[23] Meyer RE, Ezekiel S, Stowe DW, Tekippe VJ. Passive fiber-optic ring resonator for rotation sensing. Optics Letters. 1983;**8**(12):644-646

[24] Armenise MN, Ciminelli C, Dell'Olio F, Passaro VMN. Advances in Gyroscope Technologies. Heidelberg: Springer-Verlag; 2010

[25] Ciminelli C, Dell'Olio F, Armenise MN. High-Q spiral resonator for optical gyroscope applications: Numerical and experimental investigation. IEEE Photonics Journal. 2012;**4**:1844-1854

[26] IEEE Standard Specification Format Guide and Test Procedure for Single-Axis Interferometric Fiber Optic Gyros. IEEE Std 952-1997, USA; 1997

[27] Korkishko Y, Fedorov V, Prilutskiy VE, Ponomarev VG, Morev IV, Obuhovich DV, et al. Highest bias stability fiber-optic gyroscope SRS-5000. In: 2017 DGON Inertial Sensors and Systems (ISS), Karlsruhe. 2017. pp. 1-23

[28] Available at: http://www.optolink.ru/en/products/single_axis_fog

[29] Fizoptika. Available online: https://fizoptika.com/ [Accessed: 21 April 2019]

[30] Sanders GA et al. Development of compact resonator fiber optic gyroscopes. In: 2017 IEEE International Symposium on Inertial Sensors and Systems (INERTIAL), Kauai, HI. 2017. pp. 168-170

[31] Li H, Liu L, Lin Z, Wang Q, Wang X, Feng L. Double closed-loop control of integrated optical resonance gyroscope with mean-square exponential stability. Optics Express. 2018;**26**:1145-1160

[32] Ciminelli C, Peluso F, Armenise MN. A new integrated optical angular velocity sensor. Proceedings of SPIE. 2005;**5728**:93-100

Chapter 3

On the Development and Application of FOG

Xuejuan Lin, Wenlong Han, Ke Chen and Wen Zhang

Abstract

Gyroscope is a type of angular velocity measuring device, which can precisely determine the orientation of moving objects. It was first employed in navigation and later became an inertial navigation instrument widely used in modern aviation, aerospace, and national defense industries. As a vital representative of gyroscope, the fiber-optic gyroscope (FOG) has advantages in terms of compact structure, high precision, high sensitivity, and high environmental adaptability. FOG has been broadly utilized in many fields, and is also a key component of modern navigation instruments. In this paper, the history, classification, performance indicators, and application requirements of gyroscope are briefly summarized. The development history of FOG based on Sagnac effect is described in detail. The three generations of FOG are interferometric FOG, resonant FOG, and stimulated Brillouin scattering FOG. At the same time, this chapter summarizes the development and research situation of FOG in the United States, Japan, France, and other major developing countries, and compares the application of FOG in various international companies.

Keywords: gyroscope, fiber-optic gyroscope, navigation, survey

1. Gyroscope

1.1 Development of gyroscope

At the beginning of the eighteenth century, human beings discovered that the rigid body with fast rotation has fixed axis and precession. Nowadays, gyroscope is generally used to measure angular velocity and displacement in relative inertial space. In 1765, Leonhard Euler, a Russian mathematician and physicist, published the article "The Theory of Rigid Body Moving around Fixed Point," and established the basic mechanics theory of Rotor Gyroscope. Then, the dynamic equation of rigid body rotation was deduced, which laid a solid foundation for the study of gyroscope theory. In 1778, Lagrange, a French scientist, established the differential equations of motion of a rigid body rotating at a fixed point under the action of gravitational moment in his book "Analytical Mechanics." In 1852, French physicist Foucault, based on the theory of rigid body motion put forward by predecessors, combined with his in-depth study of rigid body, first discovered that the rotor rotating at high speed in the middle of the earth always pointed to a fixed direction because of inertia, and created a measuring device for verifying the rotation of the earth [1], which was named as Gyroscope. This creates a precedent for the research and development of engineering practical gyroscopes [2]. H. Anschutz and Sperry produced Gyrocompasses which were mainly applied in navigation for ships at sea [1] in 1908 and 1909. The emergence of gyrocompass marks the formation of gyroscope technology and the opening of its modern application, which pushes the theoretical research of gyroscope to practical application.

1.2 Classification of gyroscope

According to the working principle of gyroscope, it can be divided into classical mechanics-based gyroscope and modern physics-based gyroscope.

Based on the different medium of angular velocity of sensitive carrier relative to inertial space, gyroscopes can be categorized into rotor gyroscopes, optical gyroscopes, magnetohydrodynamic gyroscopes, and atomic gyroscopes in engineering. The most common rotor gyroscopes include liquid-floated gyroscopes, dynamically tuned gyroscopes, electrostatic gyroscopes, and vibration gyroscopes. Optical gyroscopes include laser gyroscopes, fiber-optic gyroscopes, and micromachined gyroscopes include MEMS gyroscopes that have been applied in engineering [3].

1.3 Performance indicators and application requirements of gyroscope

In order to analyze and evaluate the overall performance of gyroscope, a series of criteria should be formulated to provide reference for its application. Generally speaking, the main indicators of gyroscope performance are scale factor

Performance indicator	Strategic level	Inertial navigation level	Tactical level	Commercial level
Scale factor stability/ppm	<1	1–100	100–1000	>1000
Drift stability/$(°)\cdot h^{-1}$	<0.01	0.01–0.15	0.15–15	>15
Random walk/$(°)\cdot h^{-1}$	<0.01	0.01–0.05	0.05–0.5	>0.5
Range/$(°)\cdot s^{-1}$	>500	>500	>400	50~1000
Cost/$	20,000	10,000	1000	500

Table 1.
The classification of performance indicator in gyroscope.

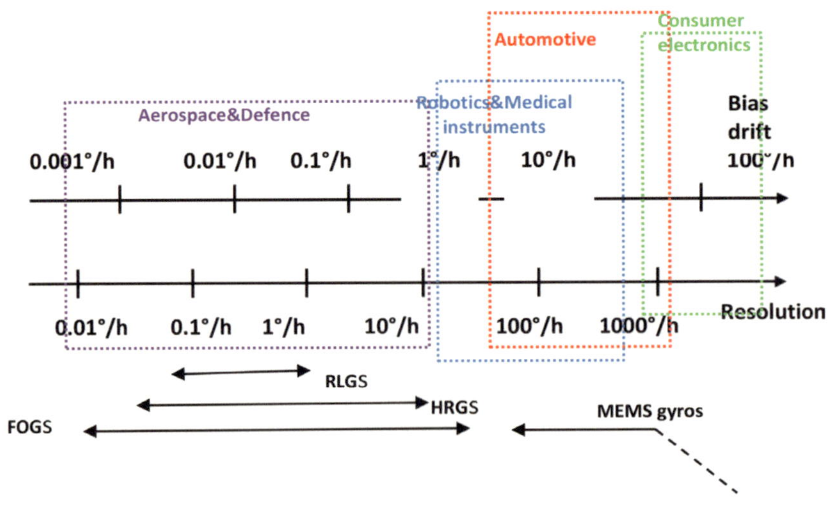

Figure 1.
The application requirements of different FOGs.

stability, drift stability, random walk, range and cost, etc. According to these indicators, gyroscopes are divided into four categories: strategic level, inertial navigation level, tactical level, and commercial level, as shown in **Table 1** [3].

Figure 1 presents the applications and requirements of different gyroscope technologies. Almost half of the high-performance gyroscope market is covered by national defense applications, while commercial aviation accounts for 25% of the market. At present, there are mainly two mature optoelectronic technologies in these two market areas, namely ring laser gyroscope (RLG) based on Sagnac effect [4] and fiber-optic gyroscopes (FOG) [5].

2. Fiber-optic gyroscope

In the 1970s, FOG was first proposed and studied [6]. Then, its emergence has opened the way for the research of all solid-state sensors. It was initially considered to be devoted to medium-level applications. But over time, it has made a number of outstanding achievements in theoretical research and engineering [7]. Nowadays, FOG has reached the strategic level of performance and surpassed the ring laser gyroscope in terms of deviation noise and long-term stability. Its advantages are becoming more obvious, and its application fields are becoming more extensive. It has gradually become the key development goal for each country.

2.1 Development history of FOG

In 1913, French physicist G. Sagnac presented a new theory through a considerable amount of experiments. The phase shift of two beams propagating along the closed optical path is proportional to the normal input angular rate of the closed optical path. That is the Sagnac effect [8]. Successful application of inertial navigation technology during World War II made FOG more challenging. In the early inertial navigation system, the sensor system used stable platform. With the progress of science and technology and the emergence of artificial satellite, people put forward the concept of strapdown inertial navigation, which has the characteristics of simple structure, small size, light weight, low cost, and easy maintenance. Sensitive devices are becoming more and more demanding. After World War II, gyroscopic technology has developed rapidly. In 1963, SePoy Gyroscope Company made a breakthrough in the area of optical gyroscope. The first experiment demonstrated ring laser gyroscope. Thereafter, after nearly 20 years of efforts, the inertial ring laser gyroscope has become practical. In 1983, Honeywell's ring laser gyroscope was installed in the airborne strapdown inertial navigation system of the new passenger aircraft Boeing 767 and 757. The rapid development of optical fiber communication, fiber optics, and laser technology has promoted the further development of optical rotation sensor based on Sagnac interferometer. In the mid- and late 1970s, a new type of optical gyroscope, named fiber optic gyroscope, appeared.

Scientists Macek and Davis confirmed the correctness and realizability of ring laser gyroscopes in 1963. In 1967, the French physicists G. Pincher and G. Herpner proposed the hypothesis of using optical fibers in gyroscopes [4]. In 1976, American scientists Victor Vali and Richard W. Shorthill tested the hypothesis of G. Pincher and G. Herpner, which symbolized the transition from theoretical stage to practical stage of FOG [9]. In 1978, McDonald Company developed the first practical FOG, and in 1980 Bergh et al. produced the first all-fiber optic gyroscope test prototype, making FOG a big step toward practicality [10]. In the mid-1980s, the interferometric fiber optic gyroscope was successfully developed. The development and application of optical gyroscope is an important milestone in the history of inertial navigation technology. FOG has great value in the military field, because of its remarkable advantages, flexible structure, and broad application prospects. It has attracted the attention of universities and scientific research institutions in many countries in the world, and has invested a lot of energy in research. At the end of 1980s and the beginning of 1990s, FOG technology has been widely used. Its sensitivity has been improved by four orders of magnitude, and the angular velocity measurement accuracy has been improved from the initial 15°/h to 0.001°/h.

2.2 Classification of FOG

The development of FOG can be roughly divided into three generations: interferometric FOG, resonant FOG, and stimulated Brillouin scattering FOG [11], as shown in **Table 2**.

2.3 Basic composition of FOG

FOG is based on solid-state technology of optical fiber communication. Specifically, the main components of FOG are shown in **Figure 2** [15]:

1. AAA, an advanced broadband source based on EDFA technology, has a wavelength of 1550 nm. Wavelength stability can be obtained by internal spectral filtering with fiber Bragg grating.

2. Polarization-maintaining optical fiber coils (hundreds of meters in mid-range and kilometers in high-grade).

3. The integrated optical circuit of Linbo3 with electrodes is used to generate phase modulation and provide good polarization selectivity through proton exchange waveguide.

4. An optical fiber coupler (or circulator for higher return power) for transmitting signals to the detector light returned from the common input-output port of the interferometer.

5. Analog-to-digital (A/D) converter for sampling detector signals.

6. The digital logic electronic device that generates phase modulation and phase feedback through a digital-to-analog converter.

Note that with proper design and components, FOG performance is repeatable in production, even for high-performance terminals.

2.4 Principle of FOG

2.4.1 Interferometric FOG

When the whole system rotates, two beams of light propagating in the opposite direction produce phase difference, and the interference intensity changes. Interferometric FOG can calculate the rotation angular velocity according to the intensity change detected by the optical detector.

The light emitted by the light source is divided into two identical beams through the beam splitter, which propagate in a closed optical path counterclockwise and

	The first generation	The second generation	The third generation
Name	Interferometric FOG	Resonant FOG	Stimulated Brillouin scattering FOG
Index	Zero-biased stability: (°/h): 8.5129 angle random walk coefficient (ARW) is 0.0841°/h~(1/2)	Zero-biased stability: (°/h): 18.181 angle random walk coefficient (ARW) is 0.05781°/h~(1/2)	The response of threshold power of pump light to temperature is $32.6 \times 10^{-6\circ}C^{-1}$, and the response of beat frequency to temperature is $88.232.6 \times 10^{-6\circ}C^{-1}$
Main features	The SAGNAC effect is enhanced by using multi-turn fiber coils. A double-beam ring interferometer consisting of multi-turn single-mode fiber coils can provide high accuracy and will inevitably make the overall structure more complex	Ring resonator is used to enhance SAGNAC effect and cycle propagation is used to improve accuracy. Therefore, shorter optical fibers can be used	Conversion of light power into light wave by stimulated Brillouin scattering
Specific classification	DepolarizedI-FOG ALLPM-fiberI-FOG IOCI-FOG	ALLfiberR-FOG IOCR-FOG(MOG)	SBS-FOG(B-FOG)
Sample grap	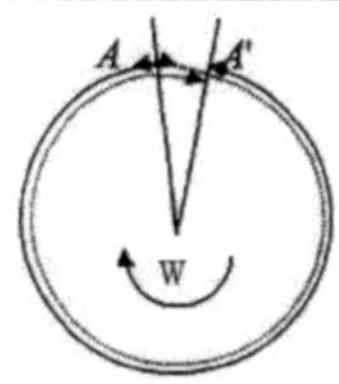	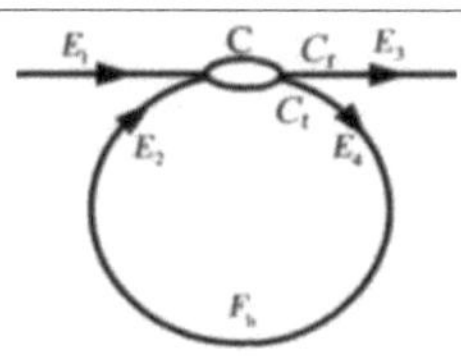	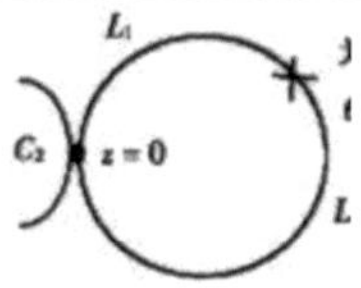
Stage of development	Practical stage	Stage of transition from laboratory to practice	Theoretical stage
Application area	Aircraft and vehicle navigation, missile guidance, precision space vehicle, submarine [11]	—	—
Advantages	Low random walk, long life, high reliability, no mechanical vibration, anti-electromagnetic interference, light weight, small size, high sensitivity, wide bandwidth, easy to realize multi-channel or distributed sensors [12]	Compared with I-FOG, the theoretical accuracy is more accurate and the volume is smaller	Simple structure, few parts, firm and stable, strong shock resistance and acceleration resistance, long service life, high sensitivity and resolution, instantaneous start-up in principle and wide dynamic range [13]
Shortcomings	Optical fibers are greatly affected by temperature. As the length increases, the cost becomes more and more expensive, the accuracy cannot be improved, and the miniaturization cannot be achieved [14]	The production cannot meet the current demand	Generation and stable output of single-frequency SBS laser, locking phenomenon, polarization fluctuation, temperature effect

Table 2.
The classification of FOG.

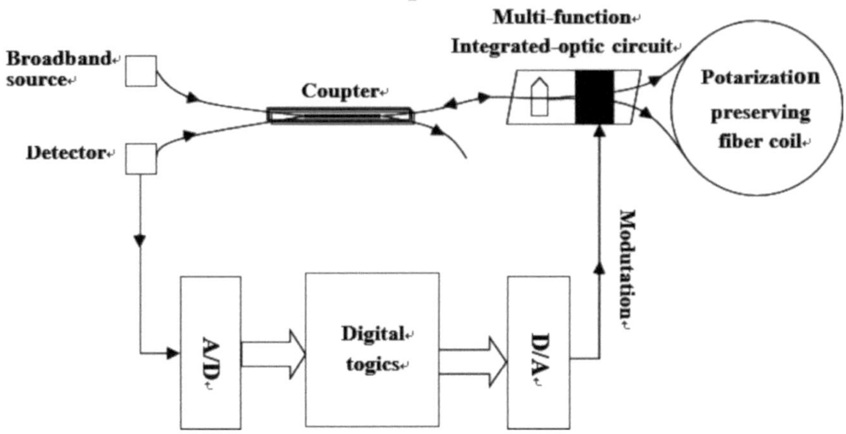

Figure 2.
The basic composition of FOG.

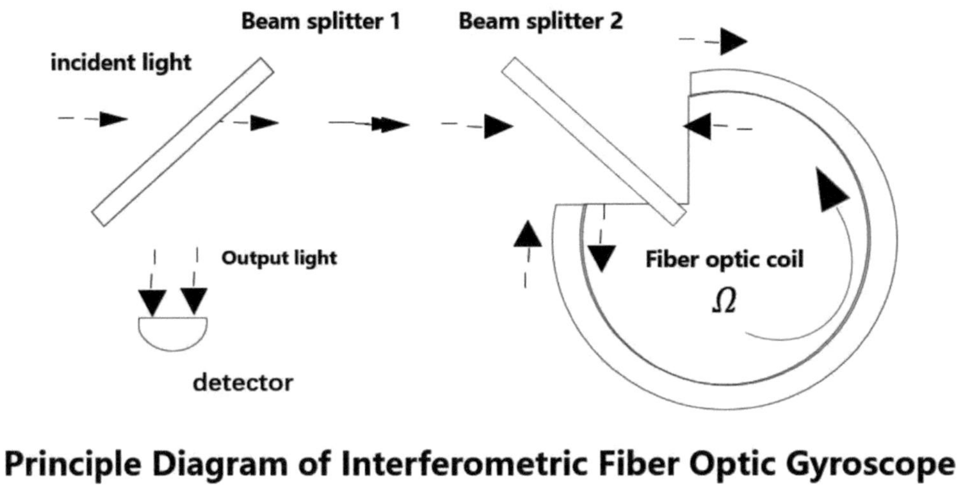

Figure 3.
The principle of interferometric FOG.

clockwise, respectively. The two beams will interfere at the beam splitter. If the closed optical path does not rotate relative to the inertial space, the two beams pass through the same path and the phase difference is zero. If the closed optical path has a rotational angular velocity relative to the inertial space, the two beams experience different paths with a slight optical path difference. At the same time, the two beams also have a phase difference, which is the Sagnac effect. IFOG uses Sagnac effect to measure rotation angular velocity. The interferometric FOG is actually the Sagnac interferometer [16]. Its schematic diagram is shown in **Figure 3**.

2.4.2 Resonant FOG

The basic principle of resonant FOG is Sagnac effect. The core device of resonant FOG is fiber ring resonator. The limit sensitivity of resonant FOG is determined by the shot noise of photodetector, so it is closely related to the resonant characteristics of resonant cavity.

Resonant Fiber Optic Gyroscope (RFOG) is also based on the clockwise and counterclockwise optical path changes caused by Sagnac effect. The light wave propagates in the optical fiber loop with periodic interference. The light source with narrow linewidth has the characteristics of long coherence, which results in resonance effect. The change of optical path of light wave propagating in optical fiber loop will lead to the change of resonance frequency point. The corresponding angular velocity can be obtained by obtaining the change of the resonance frequency point of the light wave in a certain direction.

RFOG is divided into two types, including reflective and transmission ring resonators. As shown in **Figure 4**, the reflective type uses the reflection spectrum of the resonator to detect dark peaks, while the transmission type uses the transmission spectrum of the resonator to detect bright peaks [17].

The basic principle of RFOG is Sagnac effect. For resonant gyroscope, its output detects the frequency difference of clockwise and counterclockwise beams propagating in the resonant cavity. Because it is sensitive to Sagnac frequency shift by using the steep resonant curve of the resonant cavity, it greatly reduces the length of the sensitive fiber optic coil. When the resonant cavity is stationary, the frequency difference between the two beams is zero. When the resonator rotates, the frequency of two beams of light propagating in opposite directions changes, and the frequency difference of the two beams is linear with the rotational speed. It is this frequency difference signal that the resonator gyroscope detects. Its expression is [18, 19]

$$\Delta_v = v_{ccw} - v_{cw} = \frac{4A}{\lambda L}\Omega = \frac{D}{\lambda}\Omega \tag{1}$$

where A is the area of the resonator, D is the diameter of the resonator, and 4A/λL or D/λ is the scale factor of the gyroscope. Therefore, as long as $\Delta\nu$ is measured, the rotation rate Ψ can be learned.

2.4.3 Stimulated Brillouin scattering FOG

Stimulated Brillouin scattering occurs when the intensity of the transmitted light in the fiber ring reaches threshold level. The frequency of the scattered light varies with the rotation angular velocity of the fiber ring due to the influence of Sagnac effect. The rotation angular velocity of the optical fiber ring can be obtained by detecting the frequency of the scattered light produced by CW and CCW light and beating the frequency.

Stimulated Brillouin FOG is a gyroscope consisting of Brillouin laser. It is an optical product of Ring Laser Gyroscope (RLG). Its basic principle is shown in **Figure 5**. When the incident light intensity exceeds the Brillouin threshold of the optical fiber, due to the electrostrictive effect, a moving acoustic wave will be generated

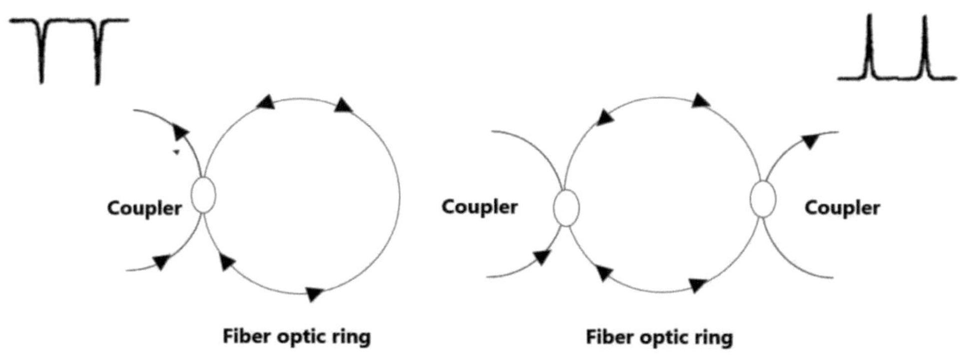

Figure 4.
The reflective resonator.

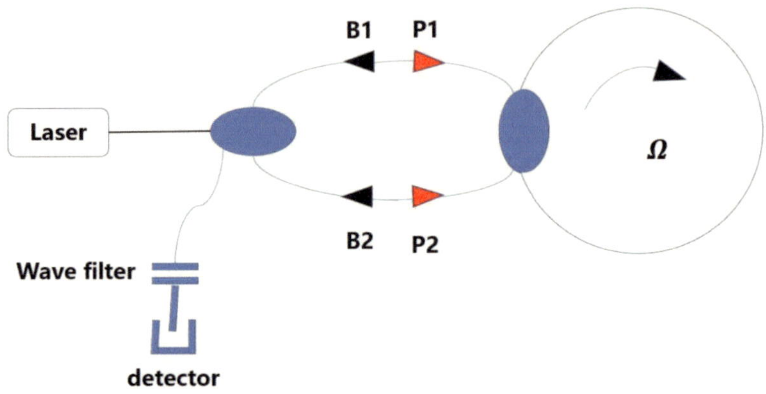

Figure 5.
The principle diagram of stimulated Brillouin FOG.

in the optical fiber. The existence of this moving acoustic wave leads to the generation of stimulated Brillouin scattering (SBS). When two pumped beams (P1 and P2) are incident into the ring resonator in the opposite direction at the same time, two Brillouin beams (B1 and B2) opposite to the pumped beams will be generated. If the ring resonator is stationary, the two Brillouin beams are proportional to the frequency difference, Dn. Two Brillouin beams are photosynthesized and beat frequency is generated. The rotation rate of the optical fiber resonator can be obtained by measuring the beat frequency, Dn.

The rotation angular velocity of the optical fiber coil is linearly related to the frequency difference of the output two Brillouin beams. The ratio factor is $\frac{4S}{\lambda \cdot nL}$, where λ is the wavelength of the pumped light, L is the length of the optical fiber coil, n is the refractive index of the optical fiber coil, and the area S is the area surrounded by the optical fiber coil.

2.5 Characteristics of FOG

1. All solid-state integration, the instrument is firm and stable, and has strong shock resistance and acceleration resistance.

2. The optical path is increased by the optical fiber ring, and the detection sensitivity and resolution are increased by several orders of magnitude compared with the laser gyroscope. Thus, the locking problem of the gyroscope is effectively overcome.

3. Without mechanical moving parts, there is no wear and tear problem, so it has a long service life.

4. The propagation time of coherent beams is very short and can start instantaneously in theory.

5. It is easy to adopt integrated optical technology. The signal is stable and reliable. It can be output digitally and connected directly with the computer interface.

6. It has a wide dynamic range.

7. It has simple structure, low price, small volume, and light weight [15].

2.6 Key technological breakthroughs of FOG

Although FOG has many advantages over other gyroscopes, it still has some shortcomings because of the imperfect technology. Thus, we can employ some solution showed in **Table 3** to obtain better performance for FOG.

2.7 State of the art of FOG

FOG has different development and research status in different countries and has its own characteristics. The United States, Japan, France, Germany, Britain, and China are the main developing countries of FOG. Europe and the United States have obvious advantages in the research and development of high-precision FOG, while Japan pays more attention to the commercial application of low-precision FOG [13]. China and other countries also attach great importance to the research and promotion of FOG.

The United States is a pioneer in developing and applying FOG. Its contractors, universities, and government agencies are developing key technologies, such as Litton, Honeywell, KVH, Norhrop Grumman, and Draper Laboratory. These companies are mainly engaged in the research and development of high-precision FOG [23], providing services for the U.S. military and aerospace departments. They have also done very well in the development and production of FOG. At present, many types of FOG have been put into use in the United States.

Japan is also a big country in the research and production of FOG. The research institutes include the cutting-edge technology laboratory of Tokyo University, Hitachi Corporation, Mitsubishi Corporation [13], Japan Aerospace Electronics Company (JAE), Mitsubishi Precision Instrument, and so on. These companies

Technical direction	Causes	Influence factor	Solution
Angular random walk coefficient (noise measurement conditions)	Back Rayleigh scattering in optical fibers and back scattering from optical interfaces	Relative intensity noise of light source, thermal phase noise of optical fiber coil and photodetector noise	Noise filtering technology; noise elimination technology [20].
Scaling temperature compensation	Important devices are sensitive to temperature	Average wavelength of light source	Broadband Erbium doped fiber light source (SFS) [21] with better wavelength stability or wavelength control measures
		Feedback channel gain	The second closed-loop feedback control loop is added based on the closed-loop feedback control circuit of FOG by using error signal [22]
Environmental adaptability	Vibration, shock, acceleration, etc.	Poor environmental adaptability	Expanding the dynamic range of measuring rotation velocity
Improving sensitivity and accuracy of detection	Poor performance of functional components	Low sensitivity and accuracy	Improving matching and phase shift of functional components [12]

Table 3.
The key technological breakthroughs of FOG.

Country	Company	Main performance: Zero bias stability	Application	Reference
America	Litton Industries Inc.	0.008°/h	The SCIT experimental inertial device was developed in 1988. Then the CIGIF demonstration system flight test device, inertial measurement system and GPS/INS integrated navigation system are developed	[25]
America	Honeywell International Inc.	0.00023°/h	It studies high-performance interferometric FOG, whose products are widely used in satellite, rocket, aircraft and other aerospace fields	[26–28]
America	Northrop Grumman	<0.005°/h	Its optical fiber technology has matured in the field of low and medium precision, and has been commercialized. Its main customers are some major airlines in the United States. Its products can be used in land, sea and air fields	[26]
Japan	Hitachi, Ltd.	Low and medium precision civil products of 10°/h	Its optical fiber technology has matured in the field of low and medium precision, and has been commercialized. Its main customers are some major airlines in the United States. Its products can be used in land, sea and air fields	[29]
Russia	Fizoptika	0.05°/h	Its FOG has been commercialized. The product models are VG949, VG941B, etc.	[30]
France	EuroFOG	Serialization from 10°/h to 0.01°/h	Tri-axis scheme is adopted below 0.1°/h, and single-axis scheme is adopted at 0.01°/h	[26, 30]
France	IXSea	0.003°/h	With a number of key patents of FOG, its application fields include offshore, underwater and space applications	[26]
Germany	LITEF	<0.01°/h	Product applications cover space, air, land and water, as well as military and civilian applications. After 2003, integrated navigation system will be provided to provide position, course and attitude information for military reconnaissance vehicles	[26]
China	Beihang University	0.005°/h	It has a complete production line of FOG	[30]
China	China Aerospace Times Electronics Co. Ltd.	0.01°/h	Its product mainly used in the field of aeronautics and astronautics	[30]

Table 4.
The application of FOG in the world.

attach great importance to the practicality of FOG. They have mass-produced a variety of levels of FOG, especially those of medium and low precision. They are in the forefront of the world in practicality and can be applied to environmental protection, vehicle navigation, industrial control, and so on.

The research and development of FOG in Western European countries mainly focus on France, Italy, and Russia. These countries attach great importance to the development of military applications of FOG. These countries are mainly committed to the development of low performance FOG equipment with drift rate greater than 1 (°)/h, Navy and air force. The first generation of FOG has been put into production. For example, PHINS series FOG, which is produced by IxSea Company in France, has been applied to inertial navigation and deep-water operation. Civitanavi Systems, Italy, based on proprietary FOG technology, developed a FOG [24] for attitude stabilization and navigation of satellite launchers.

FOG has different research and application in different countries. From **Table 4**, we can see the application of FOG in different companies in the world.

3. Conclusion and expectation

After more than 30 years on research and exploration, the technology of FOG has achieved a high level. While guaranteeing the accuracy and meeting the current requirements, FOG is gradually developing in the direction of low cost, miniaturization, high reliability, and long life.

FOG has been mainly used in astronautics, including spacecraft, satellite, aircraft, etc., and it is also widely used in civil fields such as ship, automobile navigation, mine, and so on. Based on different zero bias stability, their applications are different. If the bias stability is greater than 10°/h, it can be employed in land vehicle navigation, robot attitude control, and camera or antenna stabilization device. And when the bias stability is small ranging from 0.001 to 0.01°/h, FOG can be used in aerospace inertial navigation system and navigation. Whereas, in precision spacecraft applications, the zero bias stability required for precise aiming and tracking is less than 0.001°/h [30].

FOG is a type of angular rate measurement instrument based on Sagnac effect. It has advantages of no moving parts and wearing parts, small size, light weight, large dynamic range, fast start-up, long life, low cost, impact-resistant structure, flexible design and simple production process, etc. [29, 31]. It is broadly used in inertial navigation systems such as aviation, navigation, and aerospace, and is not in the direction of high precision. Continuous development [29, 32] with the development of modern microelectronics technology, optoelectronics technology, and signal processing technology, FOG will continue to mature; its application will continue to expand. In the future, there will be a greater stage in the field of inertial measurement.

Author details

Xuejuan Lin*, Wenlong Han, Ke Chen and Wen Zhang
School of Basic Courses of Bengbu Medical College, Bengbu, Anhui, China

*Address all correspondence to: 529373432@qq.com

References

[1] Wrigley W, Hollister WM. The gyroscope: Theory and application. Science. 1965;**149**(3685):713-715

[2] Rogers RM. Applied mathematics in integrated navigation system. AIAA Education Series. 2003;**20**(7):66-67

[3] Yefei Y, Wentao S. Development status and research of gyro technology in inertial stabilization platform. Control and Guidance. 2011;**2**(2):72-78

[4] Macek WM, Davis DTM. Rotation-rate sensing with traveling-wave ring lasers. Applied Physics Letters. 1963;**2**:67-68

[5] Culshaw B. The optical fibre Sagnac interferometer: An overview of its principles and applications. Measurement Science and Technology. 2006;**17**:R1-R16

[6] Bergh RA, Lefevre HC, Shaw HJ. An overview of fiber-optic gyroscopes. Journal of Lightwave Technology. 2003;**2**(2):91-107

[7] Song N, Ma D, Yi X, et al. Research on time division multiplexing modulation approach applied in three-axis digital closed-loop fiber optic gyroscope. Optik-International Journal for Light and Electron Optics. 2010;**121**(23):2185-2189

[8] Zhiping L, Lili W, Zhiping L. A brief history of gyroscope development. Electronic Design Engineering. 2012;**20**(7):66

[9] Pircher G, Hepner G. Perfectionnements aux dispositifs du type gyro interferometrique a laser. French Patent 1.563.720; 1967

[10] Gyroscopes and IMUs for defense, Aerospace and Industrial. Yole Development Report. 2012. Available from: http://www.reportlinker.com/p01008831-summary/Gyroscopes-and-IMUs-for-Defense-Aerospace-Industrial.html

[11] Lanfang L, Gang C, Guoliang J. Basic principles and classification of fiber optic gyroscopes. Modern Defense Technology. 2007;**35**(2):59-64

[12] Sai-qi C, Dong-li Y, Jian-guo Y, Wei J, Jian Z. An overview of fiber-optic gyroscopes. Optical Fiber and Electric Cable. 2005;**6**:6-7

[13] Haibo Z, Jianye L, Jizhou L, et al. Development status of fiber-optic gyroscopes. Sensor Technology. 2005;**24**(6):1-3

[14] Jizhou L, Jianye L, Xueyuan L, et al. Performance evaluation on IFOG. Sensor Technology. 2004;**23**(9):31-34

[15] Lefevere HC. The fiber-optioscope: Challenges to become the ultimate rotation-sensing technology. Optical Fiber Technology. 2013;**19**:828-832

[16] Chen Z. Study on polarization characteristics and temperature performance of fiber optic gyroscope. Harbin University of Engineering. 2010. p. 16-19

[17] Jun Z. Research on Processing of Interferometric Fiber Optic Gyro and Resonator Fiber Optic Gyro. Zhejiang: Zhejiang University; 2018

[18] Zhenlong X. Research on key technologies of resonant photonic crystal fiber optic gyroscope. Harbin University of Engineering. 2017. p. 113-114

[19] Yabing C, Shanghong Z, RuiPing Z, et al. Interferometric fiber optic gyroscope technology and its application progress. Progress in Laser and Optoelectronics. 2003;**40**(9):54-55

[20] Rabelo RC, Carvalho RT, Blake J. SNR enhancement of intensity

noise limited FOGs. Journal of Lightwave Technology. 2000;**18**(12):2146-2150

[21] Hongyu W. Analysis and Processing of Nonstationary Random Signals. Beijing: National Defense Industry Press; 2008

[22] Na S, Feng G, Jianlong J. Scale factor and zero bias error compensation of fiber optic gyroscope. Navigation positioning is time service.2017;**4**(4):92-96

[23] Liqin W. Fiber optic gyroscope and its application. Automation and Instrumentation. 2013;(5):132-133

[24] Lianli X, Yuetao G, Shaochun C. Development and review of inertial technology abroad in 2018. Aerial Missile; 2019. pp. 1-11. [Accessed: May 23, 2019]

[25] Yao S. Research on Signal Processing and Closed-Loop Detection Technology of Fiber Optic Gyroscope. Xi'an Technological University; 2018

[26] Wei S, Feng S, Fanming L. Development and application of FOG rotary SINS. Sensors and Microsystems. 2012;**31**(11):1-2

[27] Haitao W, Winter H, Dai Y. Patent analysis of fiber optic gyroscope technology. Winged Missile. 2019;(1):87-91

[28] Jing ZY, Yong YK, Yao P, et al. The history, current situation and prospect of gyroscopes. Cruise Missile. 2018;**408**(12):92-96

[29] Qi D, Ningfang S, Xiao WX, et al. Research on high precision online automatic tracking technology of fiber optic gyro Eigen frequency. Laser Magazine. 2019;(4):31-35

[30] Hong Z. Application and development of fiber optic gyroscope. Technical Foundation of National Defense. 2010;(3):41-42

[31] Xiaodong S, Weiyang S, Zhaohui W, Kai X, Bingfu M. Modeling and research of FOG tracking angular acceleration model. Electronic Technology and Software Engineering. 2019;(5):169-171

[32] Celikel O. Construction and characterization of interferometric fiber optic gyroscope (IFOG) with erbium doped fiber amplifier (EDFA). Optical and Quantum Electronics. 2007;**39**(2):147-156

Chapter 4

Modeling of Inertial Rate Sensor Errors Using Autoregressive and Moving Average (ARMA) Models

Mundla Narasimhappa

Abstract

In this chapter, a low-cost micro electro mechanical systems (MEMS) gyroscope drift is modeled by time series model, namely, autoregressive-moving-average (ARMA). The optimality of ARMA (2, 1) model is identified by using minimum values of the Akaike information criteria (AIC). In addition, the ARMA model based Sage-Husa adaptive fading Kalman filter algorithm (SHAFKF) is proposed for minimizing the drift and random noise of MEMS gyroscope signal. The suggested algorithm is explained in two stages: (i) an adaptive transitive factor (a_1) is introduced into a predicted state error covariance for adaption. (ii) The measurement noise covariance matrix is updated by another transitive factor (a_2). The proposed algorithm is applied to MEMS gyroscope signals for reducing the drift and random noise in a static condition at room temperature. The Allan variance (AV) analysis is used to identify and quantify the random noise sources of MEMS gyro signal. The performance of the suggested algorithm is analyzed using AV for static signal. The experimental results demonstrate that the proposed algorithm performs better than CKF and a single transitive factor based adaptive SHFKF algorithm for reducing the drift and random noise in the static condition.

Keywords: strap down inertial navigation system (SINS), MEMS gyro (MEMS), random drift, Sage-Husa adaptive Kalman filter (SHAKF), Allan variance

1. Introduction

During the last two decades, low cost, small size, accurate and reliable navigation system development is a hot research in the modern navigation technology. *The word navigation is a process of monitoring and controlling any moving object from one place to other. Inertial navigation system (INS) is a dead reckoning positioning method based on measurements and mathematical processing of the vehicle absolute acceleration and angular speed in order to estimate its attitude, speed and position related to difference* [1–5]. INS technology is categorized into (i) gimbal INS and (ii) strap-down INS. In the early 1940s, a gimbal INS system was developed based on the mechanical inertial sensor (i.e., accelerometers and gyroscopes) for providing the navigation information [5]. Its accuracy was limited by mechanical inertial sensor errors. The main drawbacks of the gimbal INS system are its designed complexity and it requires synchronous servo motors, slip rings, control electronics, etc., for acquiring the navigation information. Because of these factors, the gimbals INS systems are used in low grade navigation applications [6]. In the early 1950, strap-down INS (SINS) was developed based on solid state inertial sensor [5]. The SINS is a self-contained navigation system that has been developed for providing the accurate navigation information (i.e., position, velocity and rotation information. It has three gyroscopes ad three accelerometers. In general, the operation principle of SINS follows the physical laws of motion equations. It is an emerging technology as compared to gimbal INS systems and it has significant features such as easy to design, lower cost of ownership, moderate manufacturing cost and also high reliability. SINS consist of an inertial measurement that includes 3-axis accelerometers and 3-axis gyroscopes, and a processing computer. IMU is a key device to the INS and has been widely used for measuring the rotation rate and acceleration of an object. In practice, SINS accuracy degrades due to internal and external errors of the inertial sensors. These errors are mainly caused due to fluctuation in temperature, pressure and internal electronics components of the sensor. Due to these factors, stochastic errors and drift errors are generated at the IMU output [7, 8].

With the recent development of modern navigation technology, inertial sensor based SINS technology have been characterized into three categories, (i) low accuracy (tactical applications), (ii) medium accuracy (navigation applications) and (iii) high accuracy (strategic navigation applications) sensor technology. The performance improvements of inertial sensors are decided by the inertial sensor errors [9]. Currently, the strap-down INS use (i) low-cost MEMS and (ii) precision fiber optic gyroscope. MEMS sensor has more attractive to manufacturers of navigation systems because of their small size, low cost, light weight, low power consumption and ruggedness [10]. However, MEMS sensors give poor performance in the highly dynamic environment. Hence, the reliability of MEMS-based INS navigation accuracy is limited. Because of these features MEMS have only been used for low-end navigation applications (i.e., commercial domain) [11].

In the recent years, MEMS devices have been developed and tested successfully for low-end accuracy applications [12, 13]. MEMS sensor operates for a long time under poor condition and it generates the noise due to internal circuits and electronics interferences of the MEMS sensor [14–16]. As a result, noise and drift are generated at the MEMS output. In general, drift error is affected by ambient temperatures and magnetic field effect [17–19]. Many studies have been reported for temperature error model of MEMS sensor to capture the temperature variation affects [20]. According to the IEEE standard specification, MEMS errors can be characterized into two categories, such as (i) deterministic errors and (ii) stochastic errors. Deterministic errors are due to scale factor errors, bias and misalignment errors [18, 19]. Several calibration methods have been developed for eliminating the bias errors, scale factor errors in the lab environments. Stochastic errors are due to quantization effect, temperature effect (random bias), random drift, and additive noise of MEM sensor. In the case of stochastic errors analysis, calibration techniques cannot be suitable because of randomness [21–24]. This chapter concentrates on random errors modeling and random noise elimination techniques. The developments of random noise suppressing methods are helpful for improving the MEMS accuracy as well as SINS accuracy. In general stochastic error includes quantitation noise (QN), bias instability (BS), angle random walk (ARW), rate random walk (RRW) and rate ramp (RR) drift. With the extension of research, random noise and bias drift are the non-negligible errors in the MEMS sensor output. In this chapter, different signal processing techniques are developed to minimize the bias drift and random noise [25].

In time domain, Allan Variance (AV) is a popular technique has been widely used to identify and quantify different random noises present in the MEMS sensor [16, 26, 27]. In literature, several noise compensation techniques such as discrete wavelet transform (DWT), empirical mode decomposition (EMD) method and Forward linear prediction (FLP) methods have been developed and applied to MEMS sensors for filtering the high-frequency noise [28, 29]. These methods are not suitable when the sensor includes the correlated noise. Kalman filter (KF) is a most popular state estimation technique that has been used for minimizing the correlated noise of the MEMS sensor [30–34]. The priori knowledge of an initial values of the process and measurement noise covariance matrix are known exactly, when the KF become an optimal. However, in practice, these noise parameters may vary with time so that the performance of the KF can be degraded and then the filter become diverge.

To solve the divergence problems, Adaptive Kalman filter technique (AKF) is a better solution. The adaptation can be based on either (i) innovation based adaptive estimation AKF (IAE-AKF) or (ii) residual based estimation AKF (RAE-AKF) and also multiple model based AKF [34, 35]. Among the other methods, adaptive KF is developed using IAE. In general, an innovation sequence is defined as the difference between true and estimated values. In the IAE-AKF method, the measurement and process noise matrices are estimated based on innovation sequence and followed by sliding average window method. In real time, the selection of window size is a critical issue. Sage-Husa Adaptive KF is another version of adaptive KF that has been developed to improve the AKF performance by introducing a time varying estimator. In the SHAKF, using a time-varying noise estimator can be helpful in estimating the statistical characteristics of the uncertainty in the measurements in real time and mitigating the filter divergence. A further study on the SHAKF is developed based on adaptive factors for improving the filter practicability and optimality [23].

An adaptive fading Kalman filter (AFKF) was proposed for compensating the effect of the uncertainty in the measurements by transitive factor to the state error covariance (P). In AFKF, the state error covariance (P) is scaled with a single transitive factor for improving the filter variance and gain correction. When it is used for complex systems, the performance of AFKF degrades because of it may not be sufficient to use a single transitive factor for estimating the covariance matrix of the filter [24]. To overcome the difficulties of single transitive factor, multiple fading factors are used in AFKF. Because of that reason, authors are developed double transitive factor based SHAFKF that adapts both predicted state error covariance (P) and measurement noise covariance matrix (R) based on the innovation sequence. Although it has been successively applied to different domains, its performance for MEMS gyroscope sensor signal is not explored. The stochastic errors of MEMS gyroscope cannot be eliminated using calibration technique. It needs to be modeled before filtering the signal. Therefore, adaptive filtering techniques have been developed for minimizing the random noise from MEMS gyroscope system. In general, auto-regressive (AR), Moving Average (MA), and Auto-Regressive and Moving Average (ARMA) and Gauss-Markov model (GM) have been used for modeling stochastic signal [17]. Among these models, ARMA is a better choice for modeling MEMS gyroscope drift errors. In general, the ARMA modeling involved three steps as (i) randomness and stationary test (ii) selection of suitable time series model and (iii) estimation of model parameters. The unit root test and inverse sequence techniques have been used for checking the stationary of the signal. The model order is obtained by using auto correlation function (ACF) and partial auto correlation function (PACF). Moreover, Akaike Information Criterion also used to check the model order. The modified Yule-Walker method is used estimate the model parameters. Once an optimal ARMA model is defined, a suitable adaptive Kalman filter can be applied to minimize the drift of inertial sensors [14, 30].

In this chapter, we developed double transitive factors based on Sage-Husa adaptive fading Kalman filter (SHAFKF), namely SHAFKF-P Adaption and SHAFKF-R adaption. In addition, ARMA model is used to model the random drift errors of MEMS sensor. ARMA model based SHAFKF algorithm is developed and applied for minimizing the bias drift and random noise in the presence of MEMS gyroscope signal. The suggested algorithm is analyzed in two stages. In the first stage, the predicted state error covariance is adapted by a transitive factor, whereas, in the second stage, another transitive factor is scaled to the measurement noise covariance matrix (R). The efficiency of the algorithm is analyzed using Allan Variance technique.

The remainder of the paper is organized as follows. Section 2, explains the theory of ARMA models for MEMS gyroscope random noise analysis. The Allan Variance method is explained in Section 3. In Section 4, Conventional and adaptive Kalman filters are discussed based on innovation sequence. Section 5 explains the proposed algorithm based on double transitive factors. Designing state space model for ARMA (2, 1) model is presented in Section 6. Experimental results and static test analysis are explained in Section 7 and also followed by conclusions in Section 8.

2. Auto regressive and moving average (ARMA) model

In literature, several time series models have been widely used in many fields such as industry, science and engineering. Among the other model, auto regressive (AR) and moving average (MA) models have been most popular and since then widely used for forecasting [14–16]. The combination of AR and MA models has been used for inertial sensors error modeling. In this chapter, stationary ARMA model is proposed for characterizing the stochastic errors of the MEMS gyroscope signals. In general, the ARMA model is a combination of weighted sum of AR and MA model. The expression for the ARMA model with an order (p, q) is defined as

$$Y_n = \sum_{i=1}^{p} \varnothing_i Y_{n-i} \sum_{j=1}^{q} \theta_j \varepsilon_{n-j} + \varepsilon_n \qquad (1)$$

where p and q are the AR and MA model orders, receptively. ε_n is a sequence of independent and identical distributed random variable. Y_n is the measured time series data of MEMS gyroscope signal. $\varnothing_1, \varnothing_2, \varnothing_3, \ldots, \varnothing_p$ and $\theta_1, \theta_2, \theta_3, \ldots, \theta_q$ are the auto regressive (AR) and moving average (MA) coefficients, respectively. The MEME gyroscope sensor raw data is used to test the normality and zero mean of the time series data of MEMS gyroscope. In general, the skewness and Kurtosis should be 0 and 1 that tells that checking the zero mean and normal distributed data of the time series data of the sensors.

2.1 Time series model selection

In the time series analysis, several methods have been developed for selecting the order of the AR, MA and ARMA order. In general, auto-correlation function (ACF) and partial ACF (PACF) are the basic methods to select the model based on the characteristics of the ACF and PCF graphs as shown in **Table 1**. From **Table 1**, we observed that both ACF and PACF are tail off. In this chapter, ARMA (p, q) is suitable for modeling the MEMS Gyroscope data.

Model order	ACF	PACF
AR(p)	Tail off	Cut of at order P
MA(q)	Cut of at order q	Tail off
ARMA(p, q)	Tail off	Tail off

Table 1.
Determining the model and order of the MEMS gyro signal.

The samples autocorrelation function (ACF) is defined as

$$ACF = g_k = \frac{\frac{1}{N}\sum_{n=1}^{N-k}\left(Y_n - \mu_y\right)\left(Y_{n+k} - \mu_y\right)}{\frac{1}{N}\sum_{n=1}^{N}\left(Y_n - \mu_y\right)^2} \tag{2}$$

and the partial autocorrelation is expressed as

$$PACF = g_{kk} = \begin{cases} g_1 & if\ k = 1 \\ \dfrac{g_k \sum_{j=1}^{k-j}\left(g_{k-1}\right)\left(g_{k-j}\right)}{1 - \sum_{j=1}^{k-j}\left(g_{k-1}\right)\left(g_k\right)} & if\ k = 2, 3, \ldots, n, \end{cases} \tag{3}$$

where k is the lag and g_k is the sample autocorrelation. The μ_y and g_{kk} are the samples mean and partial correlation at lag k.

This can also cross checked using Akaike Information Criterion (AIC) method. In this work, AIC values of the time series data are evaluated using **Table 1**. The model order is selected based on the minimum value of AIC.

The general expression of Akaike information criterion (AIC) is

$$AIC = \log\left(\Theta\left[1 + \frac{2d_{aic}}{N_{aic}}\right]\right) \tag{4}$$

where Θ denotes the estimated residual of the model. d_{aic} and N_{aic} are the model order and the number of time series observation respectively.

2.2 Model parameter estimation

Suitable model parameters are estimated by using Yule-Walker, Burg, Unconstrained Least-Squares method and Levinson-Durbin methods. In general, for large data-set analysis, Yule-Walker and Unconstrained Least-Squares method are the better estimators.

3. Allan variance analysis

Allan variance (AV) is a popular time domain method has been widely used for identifying and quantifying random errors in the presence of inertial sensor [14]. Cluster based analysis is used to develop the AV technique. In the AV analysis, the IMU raw data can be divided into clusters with specified length, "m." Let us take "n" measurements of gyroscope (ω), denote it by $\omega^{x[1]}$, $\omega^{x[2]}$, $\omega^{x[3]}$, ...$\omega^{x[n]}$. The collected MEMS sensor data is sampled at rate of fs (samples per seconds). The set of samples called as cluster and denoted as "k_c". The total number of clusters can be represented by "K," (i.e., $K = \frac{n}{m}$). The measured date of the gyroscope can be written as

$$\omega^{x[1]}, \omega^{x[2]}, \omega^{x[3]}, \ldots\ldots\omega^{x[m]}, \omega^{x[m+1]}, \ldots\ldots\omega^{x[2m]}, \ldots\ldots\omega^{x[n-m]}\ldots\omega^{x[n]}$$

To calculate each clusters average is

$$\overline{\omega}^{x[k_c]} = \sum_{i=1}^{m} \overline{w}^{x[k_c-1]m+i} \tag{5}$$

Here, k_c =1, 2, 3, ..., K is the number of clusters.

The Allan variance is computed from two successive cluster averages for the specified correlation time which is defined as:

$$\sigma^2(\tau_m) = \frac{1}{2(K-1)} \sum_{k_c=1}^{K-1} \left(\left(\overline{\omega}^{x[k_c+1]}(m) - \overline{\omega}^{x[k_c]}(m) \right)^2 \right) \tag{6}$$

where k_c =1, 2, 3, ..., K, and $\tau_m = {}^{m}/_{f_s}$ is averaged period (or specified correlation time). The AV can be computed as a function of correlation times versus the Allan deviation plot are shown in **Figure 1**. The different contribution error sources are carried out simply by examining the slope of the AV plot. To extract the information on a specific source of error from the AV plot.

There is a unique relationship between the Allan Variance (time domain) and the PSD (frequency domain) of the random process as:

$$\sigma^2(\tau_m) = 4 \int_0^{\infty} S_\Omega(f) \frac{\sin^4(\pi f T)}{(\pi f T)^2} \tag{7}$$

where $S_\Omega(f_s)$ is the power spectral density (PSD) of the random process and $\frac{\sin^4(\pi f T)}{(\pi f T)^2}$ is the transfer function of PSD.

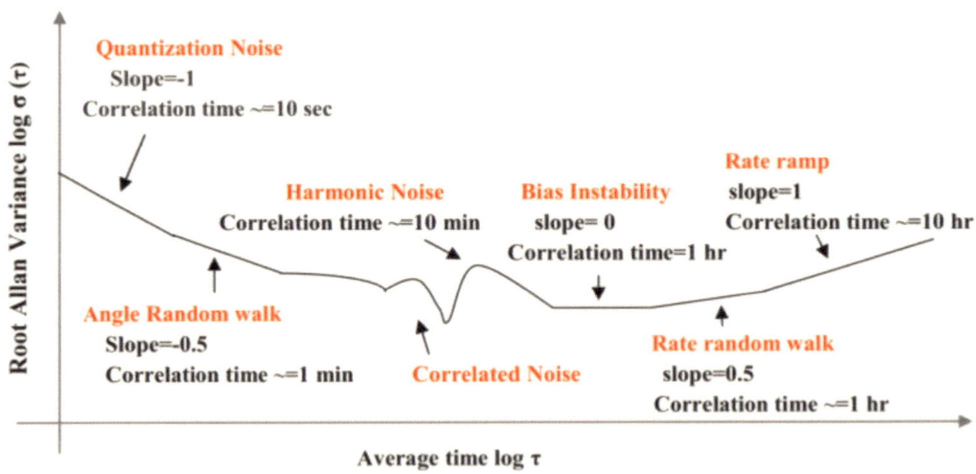

Figure 1.
Allan variance log-log plot.

Noisy type	Units	Slope	Root Allan variance
Quantization noise (QN)	$^\circ/\sec$	−1	$\sigma_{QN}(\tau) = \frac{\sqrt{3}QN}{\tau}$
Angle random walk (ARW)	$^\circ/\sqrt{hr}$	−1/2	$\sigma_{ARW}(\tau) = \frac{ARW}{\sqrt{\tau}}$
Bias instability (BS)	$^\circ/hr$	0	$\sigma_{BS}(\tau) = 0.668\ Bs$
Rate random walk (RRW)	$^\circ/\sqrt[3]{hr}$	1/2	$\sigma_{RRW}(\tau) = RRW\sqrt{\frac{\tau}{3}}$
Rate ramp (RR)	$^\circ/hr^2$	1	$\sigma_{RR}(\tau) = RRW\frac{\tau}{\sqrt{2}}$

Table 2.
Allan variance analysis results.

The different random noise processes are characterized at various frequencies that are fitted by the AV method. The root Allan variance with each correlation time and slope are computed and presented in **Table 2**.

4. Adaptive Kalman filtering

4.1 Conventional Kalman filter

The application of conventional Kalman filter (CKF) for the MEMS gyroscope requires a prior knowledge of dynamic process and measurement models. In addition, the process and measurement noise of the MEMS gyroscope. Considering a linear dynamic system, the state and measurement equations can be written as

$$x_k = Ax_{k-1} + Bu_k + w_k \tag{8}$$

$$z_k = Hx_k + v_k \tag{9}$$

where x_k is the state vector at epoch k; A is the state transition matrix; w_k is the system (process) noise; z_k is the observation (measurement) at epoch k; H represents the observation matrix; and v_k is the measurement noise. Let us assume that the process w_k and measurement noises (v_k) are the Gaussian white noise with zero mean and finite variance that means that these are statistically independent from each other, following properties can be satisfied:

$$\mathrm{E}\ \{w_k\} = 0, \mathrm{E}\ \{v_k\} = 0 \tag{10}$$

$$\mathrm{E}\ \left\{w_k w_k^T\right\} = Q_k \tag{11}$$

$$\mathrm{E}\ \left\{v_k v_k^T\right\} = R_k \tag{12}$$

Basically, the Kalman Filtering estimation algorithm comprises two steps, namely prediction and updating equations. The main Kalman Filtering equations are given below.

Prediction equations can be expressed as

$$\hat{x}_k^- = A\hat{x}_{k-1} \tag{13}$$

$$P_k^- = AP_{k-1}A^T + Q_k \tag{14}$$

In the above equations, A is the state transition matrix and A^T denotes the transpose of A. P_k^- and Q_k represents prediction state error covariance and process noise covariance matrix at epoch k.

In the linear Kalman filter, the measurement updated equations are

$$K_k = P_k^- H^T \left(H P_k^- H^T + R \right)^{-1} \tag{15}$$

$$\hat{x}_k = \hat{x}_k^- + K_k \left(z_k - H \hat{x}_k^- \right) \tag{16}$$

$$P_k = (I - K_k H) P_k^- \tag{17}$$

where $\hat{x}_k$ is the estimated state, K_k is the gain matrix and P_k is the estimated of state vector. R and I are the measurement noise covariance matrix and identify matrix respectively.

4.2 Innovation based adaptive estimation adaptive Kalman filter (IAE-AKF)

CKF requires a prior knowledge of the measurement and dynamic process models of MEMS IMU. In practice, statistical noise models of the process and measurement models are varying with time because of that the CKF would deprive optimality. To address this divergence, an adaptive KF (AKF) is a better solution. In the AKF, the adaptation can be carried out using three ways: (a) varying Q by trial and error until a stable state is estimated with fixed R [20]; (b) varying R by keeping Q fixed; (c) varying Q and R simultaneously [21]. In the IAE-AKF algorithm, we selected the second adaption method is that varying the measurement noise covariance matrix (R) by keeping Q fixed based on innovation sequence V_k.

The innovation sequence is defined as the difference between true measurements and predicated measurements that can assume to be additional information to the filter. The innovation sequence is a zero-mean white Gaussian noise sequence, defined as

$$V_k = z_k - H \hat{x}_k^- \tag{18}$$

The weighted innovation $K_k \left(z_k - H \hat{x}_k^- \right)$ acts as a correction to the predicted estimation $\hat{x}_k^-$ to form $\hat{x}_k$. By substituting the measurement model (5) in (14), we get $V_k = H \left(x_k - \hat{x}_k^- \right)$. By taking variance on both sides of this, the theoretical covariance matrix of V_k is

$$C_{V_k} = H P_k^- H^T + R_k \tag{19}$$

The optimal estimation of covariance matrix of innovation sequence using average window method can be expressed as

$$\hat{C}_{V_k} = \frac{1}{D_s} \sum_{j=j0}^{D_s} V_j V_j^T \tag{20}$$

where V_j is the innovation sequence, D_s is the window size, $j0 = k - D_s + 1$ is the first epoch. If the window size is too small, the measurement estimation covariance can be noisy; on the other hand, the estimation of measurement covariance will be smoother. Usually, window size is chosen empirically for statistical smoothing.

The estimated measurement noise covariance based on innovation sequence is

$$\hat{R}_k = \hat{C}_{V_k} - HP_k^- H^T \tag{21}$$

where $\hat{R}_k$ is the estimated measurement noise covariance matrix, H is the observation matrix, P_k^- is the prediction state error covariance and $\hat{C}_{V_k}$ is the estimated covariance matrix of innovation sequence.

4.3 Sage-Husa adaptive Kalman filter (SHAKF)

Sage-Husa AKF (SHAKF) is another class of adaptive filtering that uses a time-varying noise statistical estimator to proceed recursively. It is also used to reduce the sensor noise in the presence of MEMS IMU signals [16]. The linear dynamical process and measurement model equations can be written in the Eqs. (4) and (5).

The expectation and the covariance matrices of w_k and v_k are written as.

$$\text{E}\{w_k\} = \hat{q}_k \tag{22}$$

$$\text{E}\{v_k\} = \hat{r}_k \tag{23}$$

$$\text{E}\left\{w_k w_k^T\right\} = \hat{Q}_k \tag{24}$$

$$\text{E}\left\{v_k v_k^T\right\} = \hat{R}_k \tag{25}$$

where $\hat{Q}_k$ and $\hat{R}_k$ are the initial estimated process and measurement noise covariance matrices, respectively.

The time-varying noise statistic recursive estimator is given by:

$$\hat{r}_{k+1} = (1 - d_k)\hat{r}_k + d_k\left(z_k - H\hat{x}_k^-\right) \tag{26}$$

$$\hat{R}_{k+1} = (1 - d_k)\hat{R}_k + d_k\left(V_k V_k^T - HP_k^- H^T\right) \tag{27}$$

$$\hat{q}_{k+1} = (1 - d_k)\hat{q}_k + d_k(x_k - A\hat{x}_k) \tag{28}$$

$$\hat{Q}_{k+1} = (1 - d_k)\hat{Q}_k + d_k\left(K_k V_k V_k^T K_k^T + P_k - AP_{k-1}A^T\right) \tag{29}$$

where $d_k = (1 - b_k)/\left(1 - b_k^{k+1}\right)$ is the amnestic factor, value range between 0 and 1. The Kalman filtering output signal and Sage-Husa self-adaptive Kalman filtering output signal are expressed in terms of following equations.

Prediction equations as.

$$\hat{x}_{k-1}^- = A\hat{x}_{k-1} + \hat{q}_k \tag{30}$$

$$P_{k-1}^- = AP_{k-1}A^T + \hat{Q}_{k-1} \tag{31}$$

Measurement updated equations are equations:

$$K_k = P_k^- H^T \left(HP_k^- H^T + R_k\right)^{-1} \tag{32}$$

$$\hat{x}_k = \hat{x}_k^- + K_k\left(z_k - H\hat{x}_k^-\right) \tag{33}$$

$$P_k = (I - K_k H)P_{k-1}^- \tag{34}$$

Here, an innovation sequence can be written as

$$V_k = z_k - H\hat{x}_k - \hat{r}_k \tag{35}$$

The K_k is the Kalman updated gain. R_k and I are the measurement noise covariance matrix and identity matrix respectively.

5. Proposed: Sage-Husa adaptive fading Kalman filter (SHAFKF) based on double transitive factors

Adaptive estimation methods have been developed for improving the CKF performance [22, 23]. In the AKF, covariance matching techniques is used to estimate the covariance matrix of the innovation or residual by fixing the values of Q. By using a scale factor in the AKF hence the performance filter was improved for estimating the state error covariance and also it improves the variance of the predicted state. Further, adaptive fading Kalman filters have been developed for improving the filter performance by introducing multiple adaptive scaling factors [24]. In the proposed algorithms, adaptive transitive factors based linear Adaptive Kalman filter algorithm is proposed also used for improving the MEMS gyroscope performance [25, 30]. However, a limited work has been reported the use of transitive factors in ARMA model based Sage-Husa KF. The proposed algorithm is explained in two cascaded stages. The predicted state error covariance P is adapted in the stage one, whereas in the second stage, the measurement noise covariance R is adapted by another transitive factor. The proposed scheme is shown in Sections 5.1 and 5.2, respectively.

5.1 Stage one: adaptation of predicted state error covariance (P)

In this stage, the predicted state error covariance is modified using an adaptive transitive factor. This stage is also termed as SHAFKF-P adaptation. The transitive factor is used to reduce the process noise of kinematic model based on the residual sequence.

The transitive factor $a_1(\mathrm{k})$ is evaluated as

$$a_1(\mathrm{k}) = \begin{cases} 1, & tr\left(C_{V_k}\right) > tr\left(\hat{P}_{\overline{v}k}\right) \\ \dfrac{tr\left(\hat{C}_{V_k} - R_k\right)}{tr\left(C_{V_k} - R_k\right)}, & Otherwise \end{cases} \tag{36}$$

where tr is the trace function and $\hat{P}_{\overline{v}k}$ is the estimated covariance matrix of the residual sequence expressed as

$$\hat{P}_{\overline{v}k} = \overline{V}_k \overline{V}_k^T \tag{37}$$

The predicted state covariance $\hat{P}_k^-$ is updated as

$$\hat{P}_k^- = \frac{1}{a_1(k)} \hat{P}_{k-1}^- \tag{38}$$

The SHAFKF-P adaptation algorithm, the predicted and estimated state error covariance are updated based on the SHAKF algorithm.

$$\hat{P}_k^- = \frac{1}{a_1(k)} \hat{P}_{k-1}^- \tag{39}$$

$$C_{V_k} = HP_k^- H^T + R_k \tag{40}$$

The suboptimal state and update the measurement equations as

$$K_k = P_k^- H^T \left(H P_k^- H^T + R_k \right)^{-1} \tag{41}$$

$$\hat{x}_k = \hat{x}_k^- + K_k \left(z_k - H \hat{x}_k^- \right) \tag{42}$$

$$P_k = (I - K_k H) P_{k-1}^- \tag{43}$$

where $\hat{x}_k$ indicates the suboptimal estimated state vector and C_{V_k} denotes the suboptimal covariance matrix of innovation the state vector. For optimal filter purpose, $\hat{x}_k$ and P_k are further passed to the next stage. The flowchart of the stage one of the algorithm is shown in **Figure 2**.

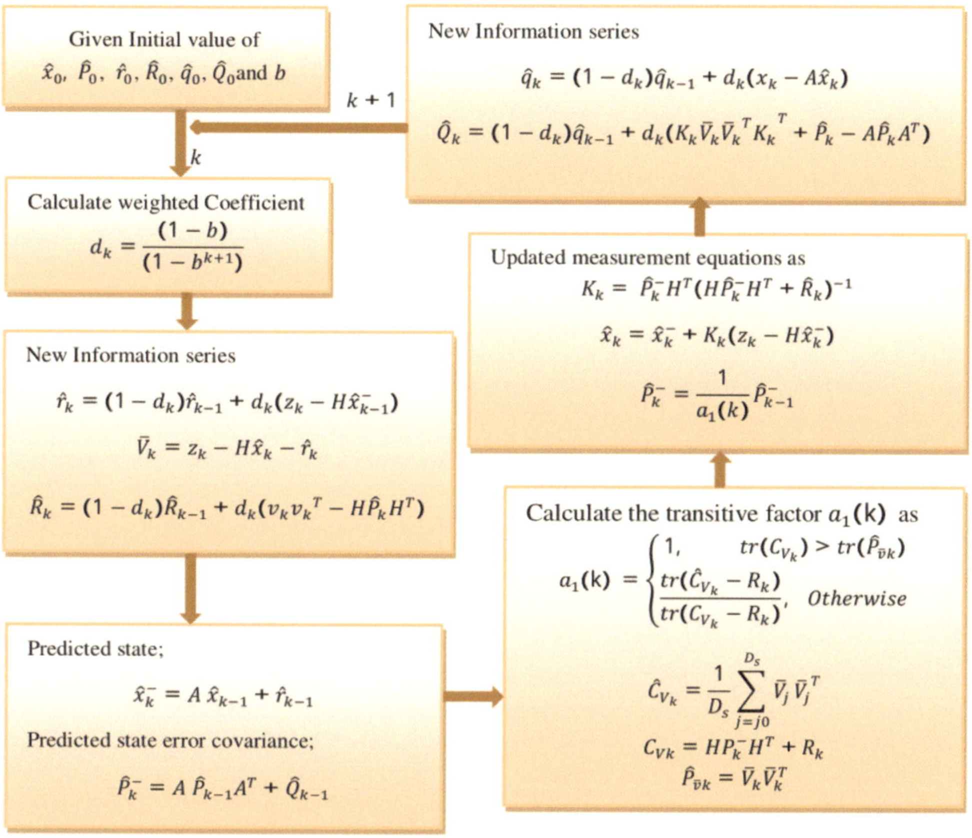

Figure 2.
Flow chart of the SHAFKF-P adaptation algorithm.

5.2 Stage two: adaptation of measurement noise covariance matrix (R)

The stage one algorithm requires prior knowledge of the state error vector and kinematic of model errors. To overcome this drawback and to eliminate the influence of the measurement noise disturbances, another transitive factor is introduced for updating the measurement noise covariance matrix (R). This stage is also termed as AUFKF-R adaptation.

In this stage, modified residual sequence is evaluated as the difference between measurement vector **zk** and the suboptimal estimated state ($\hat{x}_k$) evaluated using Eq. (45). Thus the modified residual sequence can be defined as

$$\overline{V}_k = z_k - H\hat{x}_k - \hat{r}_k \tag{44}$$

Furthermore, using the suboptimal state error covariance $\hat{P}_{\overline{v}k}$ similar to Eq. (41), the estimated covariance matrix of the residual sequence can be written as

$$C_{\overline{v}k} = HP_k^- H^T + R_k \tag{45}$$

The suboptimal estimation of covariance matrix of residual sequence using the average window method is

$$\hat{C}_{\overline{v}k} = \frac{1}{D_s}\sum_{j=j0}^{k} \overline{V}_k(j)\overline{V}_k^T(j) \tag{46}$$

The transitive factor $a_2(\mathrm{k})$ for the stage two is evaluated as

$$a_2(\mathrm{k}) = \begin{cases} 1, & tr(C_{\overline{v}k}) > tr\left(\hat{C}_{\overline{v}k}\right) \\ \dfrac{tr(C_{\overline{v}k})}{tr\left(\hat{C}_{\overline{v}k}\right)}, & Otherwise \end{cases} \tag{47}$$

In this algorithm, the measurement noise covariance matrix is scaled by a factor $a_2(\mathrm{k})$. Thus Eq. (48) can be rewritten as

$$C_{\overline{v}k} = HP_k^- H^T + a_2(\mathrm{k})R_k \tag{48}$$

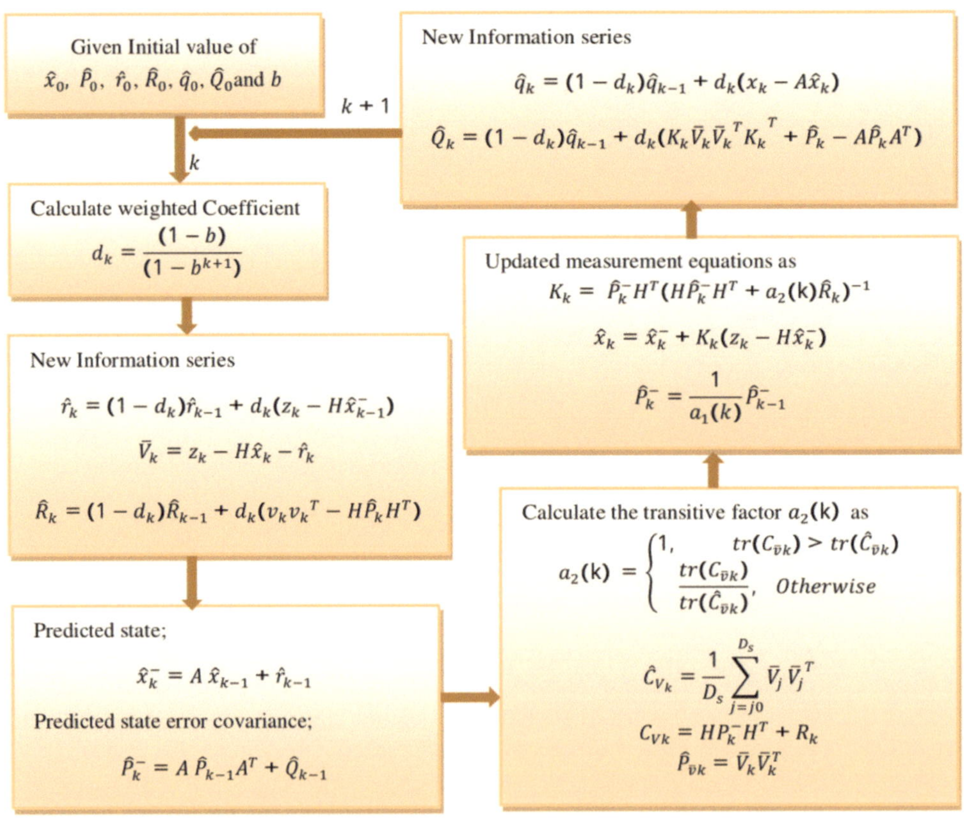

Figure 3.
Flow chart of the SHAFKF-R adaptation algorithm.

The Kalman gain and state equations are updated as Eqs. (41)–(46). In this algorithm, measurement noise covariance matrix is multiplied by the adaptive transitive factor, $a_2(k)$. If $a_2(k)$ large, R_k becomes larger, this helps to reduce the influence of uncertain measurement noise [23, 24]. The flow chart of the stage two, i.e., SHAFKF-R adaptation, is shown in **Figure 3**.

6. Designing state space model for ARMA (2, 1) model

The ARMA (p, q) model order is obtained using AIC method as in **Table 3**. The minimum values of AIC can be decided the optimal order of the ARMA (2, 1) is chosen. The ARMA (2, 1) model parameters such as **Φ1 = −0.5422, Φ2 = −0.1204** and $\theta 1$ = **0.1382** are estimated based on the minimum AIC value, i.e., −5.7612. The parameters are tabulated in **Table 3**.

The ARMA (2, 1) model is used to approximate the MEMS Gyro sensor as:

$$Y_n = \varphi_1 Y_{n-1} + \varphi_2 Y_{n-2} + \theta_1 \varepsilon_{n-1} + \varepsilon_n \tag{49}$$

where **Φ** is the AR coefficients and θ is the MA model parameter, ε_n is the system input Gaussian white noise with zero mean and variance σ_n^2. State-space representation of the optimal ARMA (2, 1) model is described as

$$X_k = \begin{bmatrix} \varphi_1 & \varphi_2 \\ 1 & 0 \end{bmatrix} \begin{bmatrix} Y_{n-1} \\ Y_{n-2} \end{bmatrix} + \begin{bmatrix} 1 & \theta_1 \\ 0 & 0 \end{bmatrix} W_k \tag{50}$$

$$Z_k = \begin{bmatrix} 1 & 0 \end{bmatrix} \begin{bmatrix} Y_n \\ Y_{n-1} \end{bmatrix} + V_k \tag{51}$$

where $W_k = \begin{bmatrix} \varepsilon_k & \varepsilon_{k-1} \end{bmatrix}^T$ is the process noise. The initialize the state estimate $\hat{x}_0 = \begin{bmatrix} 0 & 0 \end{bmatrix}^T$ and state error covariance, $\hat{P}_0 = I$ are selected. In practice, the process noise covariance matrix and the measurement noise covariance matrix are assumed as $\begin{bmatrix} \sigma_{W_k}^2 & 0 \\ 0 & \sigma_{V_k}^2 \end{bmatrix}$. In the CKF, the process and measurement noise covariance matrices are constant whereas in the adaptive proposed algorithms, these parameters are changed iteratively.

Model	φ_1	φ_2	θ_1	AIC
AR(1)	−0.5422			−5.1769
AR(2)	−0.5422	−0.1204		−5.1832
MA(1)			−0.1382	−5.1860
ARMA(1, 1)	−0.5422		−0.1382	−5.5726
ARMA(2, 1)	−0.5422		−0.1382	−5.7612

Table 3.
Parameters estimation results with AIC values.

7. Test results and discussion

The experimental setup consists of a single axis a prototype Xsens MTi 10 series MEMS sensor, turn table control unit, data acquisition board, and data processing computer. The MEMS gyroscope specification and test conditions of three single axis gyro sensor detailed results are reported in [36]. The experimental raw data is collected for 1 hour duration with sampling frequency at 100 Hz at room temperature. In the static condition, MEMS gyro is in zero rotation under the room temperature, for a more detailed specification of the Xsens MTi 100 series MEMS please refer to [36, 37].

7.1 Static performance test analysis

Three single-axis MEMS gyro sensor raw data are collected for 1 hour duration with sampling frequency at 100 Hz. The pre-processing methods are required to

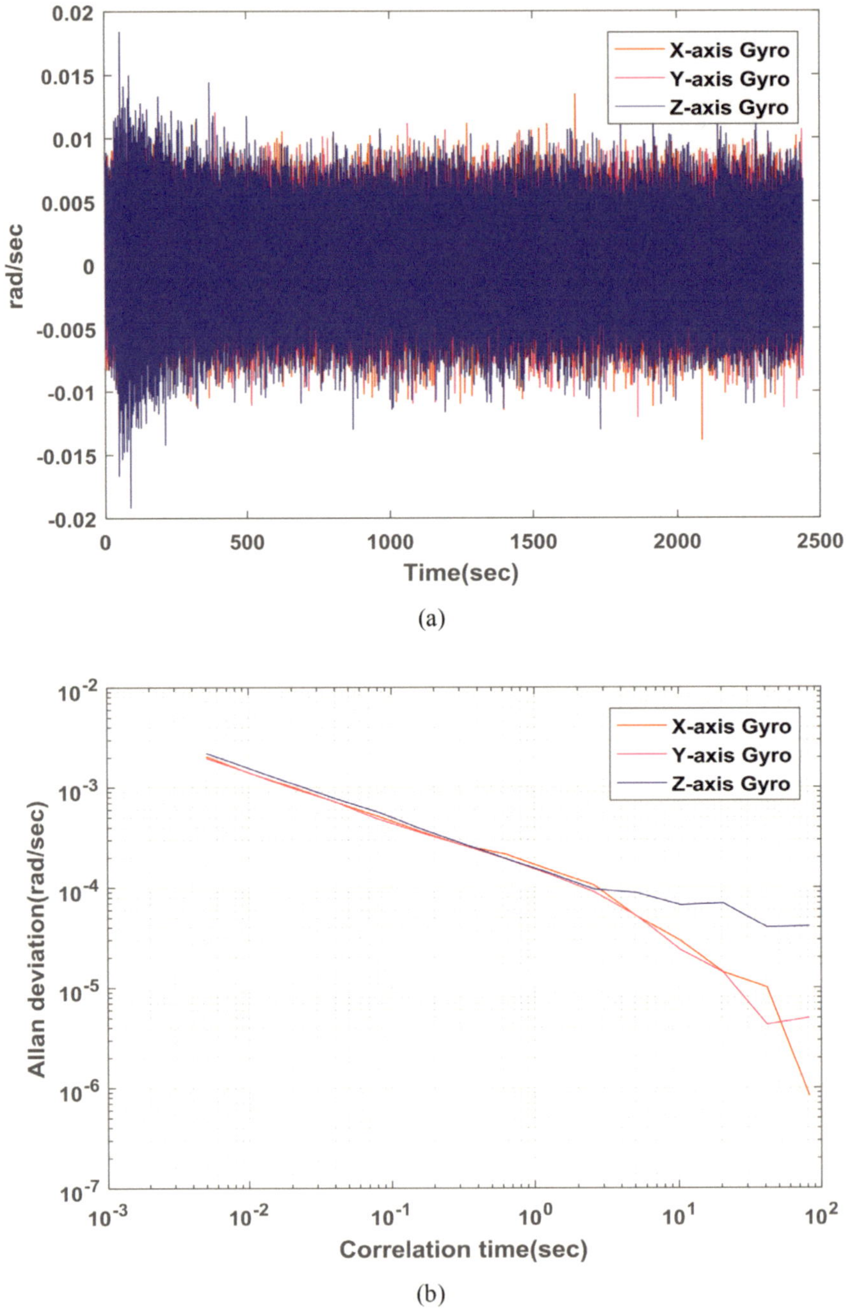

Figure 4.
(a) Three single axes of MEMS gyroscope raw signals and (b) corresponding Allan variance plot.

test the zero mean values for the sensor raw data before analyzing the Allan variance (AV) results [16]. Three single-axes of the MEMS Gyro sensor signals and corresponding AV results are plotted in **Figure 4a** and **b** respectively. From these figures, we see that the −1/2 slope indicates the angle random walk (ARW), which is a white noise characteristics. Bias instability (Bs) is due to internal and external electronic components of the sensor and is indicated at zero slope in log-log AV plot [16]. The three axes of MEMS IMU sensors are identified and quantified using AV

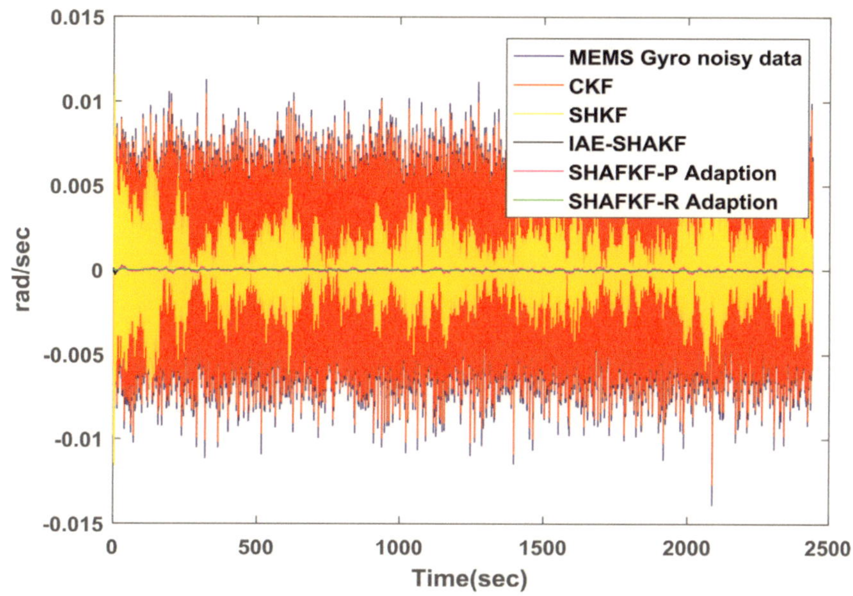

(a)

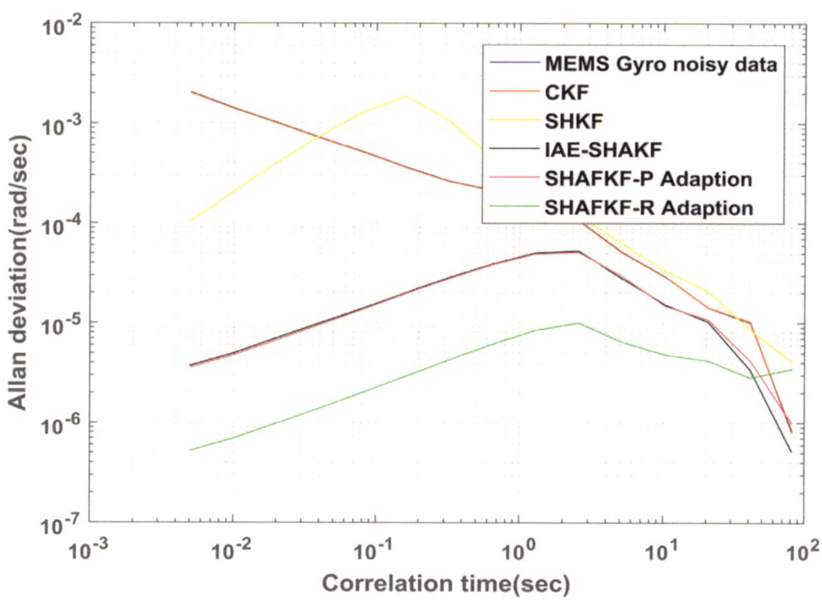

(b)

Figure 5.
(a) X-axis MEMS gyro signal and de-noised results using the SHAFKF algorithm and (b) corresponding Allan variance plot.

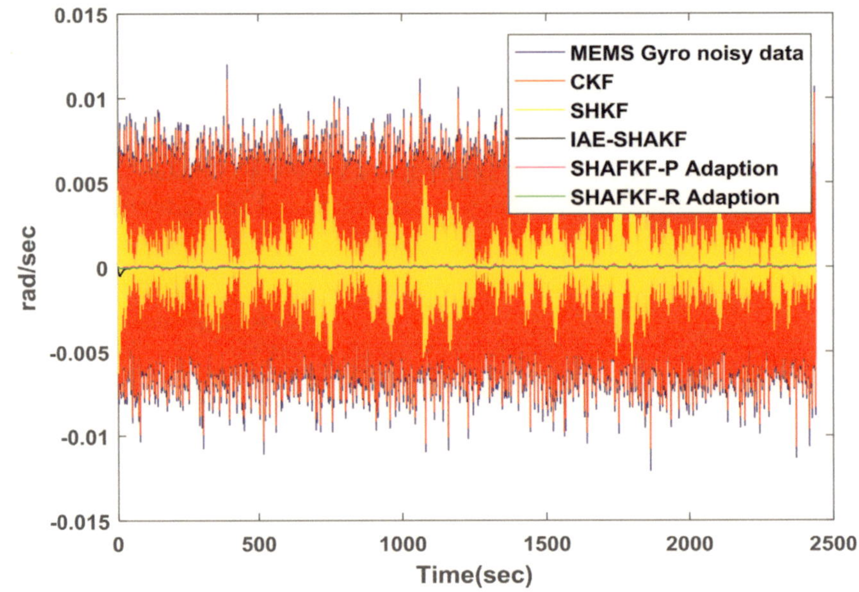

(a)

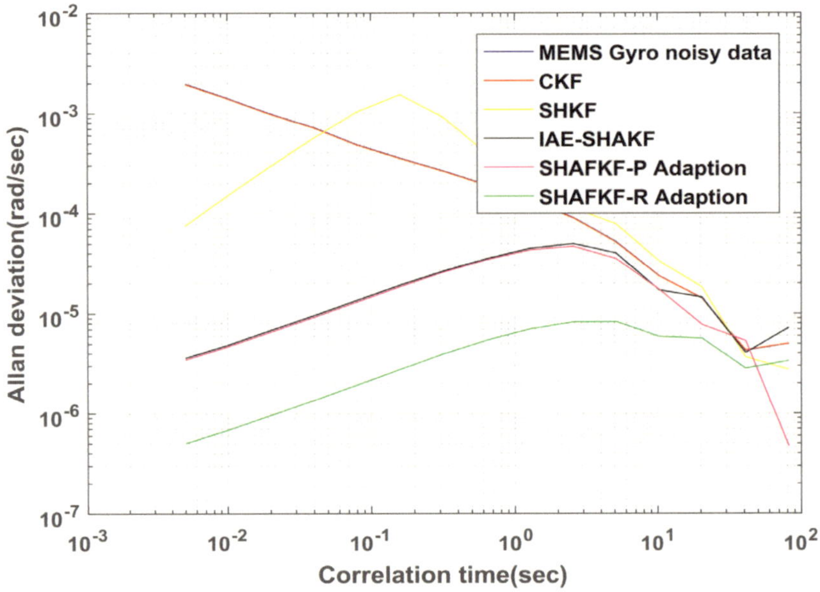

(b)

Figure 6.
(a) Y-axis MEMS gyro signal and de-noised results using the SHAFKF algorithm and (b) corresponding Allan variance plot.

analysis, which are presented in **Table 2**. From this table, we can observe that ARW and BI are the two that dominate noises in the presence of the MEMS sensor.

Conventional Kalman filter (CKF) algorithm is applied for minimizing the all three axis MEMS gyro static signal. In this experiment, the initial values of measurement and process noise covariance matrix are chosen as 0.098 and 0.0001 respectively. In practical application, these noise covariance matrices vary with time. In real-time, by adjusting the noise parameters are critical. The adaptive KF

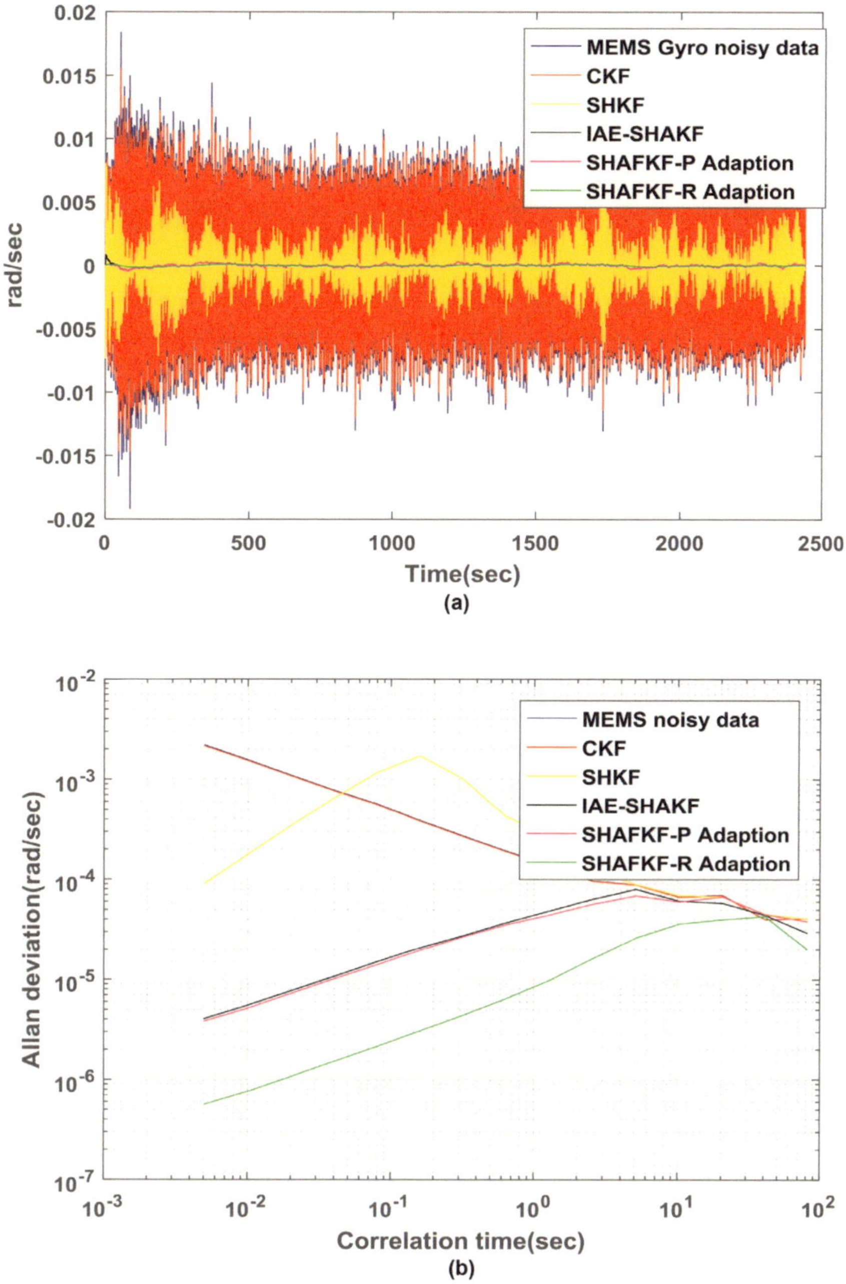

Figure 7.
(a) Z-axis MEMS gyro signal and de-noised results using the SHAFKF algorithm and (b) corresponding Allan variance plot.

algorithm, an innovation sequence is used to adjust the noise parameters of process and measurement noise matrices ad it is followed by covariance matching principle. In the IAE-AKF algorithm, the window width selection is critical and can decide the filter optimality. In general, the window width is varied between 5 and 30. In this analysis, we observed that 15 samples of the window width is the optimal choice for statistical smoothing.

In the SHAKF algorithm, the innovation sequence is used to estimate the measurement noise covariance matrix and followed by sliding window average method. In addition, statistical noise estimator is used in the AKF frame work for updating the noise coefficients in each iteration recursively. The window width is 15 samples for statistical smoothing. The SHAKF algorithm results are plotted in **Figures 5a–7a**, respectively.

In the proposed approach, the predicted state error covariance is updated by one transitive factor whereas the measurement noise covariance matrix is updated using another transitive factors based on the residual sequence. The covariance matrix of residual sequence is estimated using sliding average window method. In this method, window width is chosen empirically as 15. In the first stage of the proposed algorithm (SHAFKF-P adaption), the transitive factor (a_1) is calculated in stage one. The measurement noise covariance matrix is scaled by an adaptive transitive factor (a_2) is in the second stage. The transitive factors are used to scale R_k and reciprocal to $\hat{P}_k^-$ for reducing the variance of uncertainty in the process model and measurements, respectively. The developed algorithm is also applied to X, Y and Z-axis MEMS gyroscope static signal. The test results of the proposed algorithm for X, Y and Z-axis data are shown in **Figures 5a–7a**, respectively. From these figures, it is observed that the angle random walk (ARW) and bias instability (Bs) noise are the dominated noise sources. The quantified noise coefficients are tabulated in the **Tables 4–6**, respectively. All the random noise and drift are quantified before and after applying the de-noising algorithm. The drift is also calculated before and after de-noising MEMS signal and tabulated in **Tables 4–6**, respectively. From these tables, it is observed that the ARW is reduced by 1000 and also Bs random noise is minimalized by order of 100 compared to the original value.

From these tables, it is evident that SHAFKF of R adaptation using transitive factor improves the performance of the algorithm. In this proposed algorithm,

Methods	ARW ($^\circ/\sqrt{hr}$)	BS ($^\circ/hr$)	Drift ($^\circ/hr$)
MEMS raw data	165.115	8.775	1.758
CKF	103.235	7.459	1.362
IAE-AKF	24.858	3.496	0.859
SHAKF	4.228	2.296	0.0014
SHAFKF-P Adaption	1.279	0.690	0.00038
SHAFKF-R Adaption	0.331	0.421	0.00012

Table 4.
Allan variance and drift results of X-axis MEMS gyro using proposed scheme in static condition.

Methods	ARW ($^\circ/\sqrt{hr}$)	BS ($^\circ/hr$)	Drift ($^\circ/hr$)
MEMS raw data	33.0437	4.7297	1.7587
CKF	30.6510	1.5762	1.3624
SHKF	4.0098	1.2862	0.0859
IAE-SHAKF	3.4075	0.4371	0.00386
SHAFKF-P adaption	0.6570	0.2653	0.000562
SHAFKF-R adaption	0.442	0.150	0.000312

Table 5.
Allan variance and drift results of Y-axis MEMS gyro using proposed scheme in static condition.

Methods	ARW ($°/\sqrt{hr}$)	BS ($°/hr$)	Drift ($°/hr$)
MEMS raw data	38.9222	9.8105	1.758
CKF	24.3805	8.3228	1.216
SHKF	7.0068	3.502	0.597
IAE-SHAKF	3.3229	3.071	0.0013
SHAFKF-P adaption	1.2516	0.832	0.00028
SHAFKF-R adaption	0.914	0.542	0.00014

Table 6.
Allan variance and drift results of Z-axis MEMS gyro using proposed scheme in static condition.

measurement noise covariance is scaled by the transitive factor. It ensures the variance is inversely proportional to the uncertainty of measurement. Due to this, SHAFK-R adaptation algorithm outperforms other algorithms.

In addition, we observed the Drift error for the MEMS gyroscope signals. Drift error is considered as one of the performance indicator of all the proposed algorithms. From **Tables 4–6**, it is observed that the proposed SHAFKF-R adaptation filter performs better than CKF, IAE-AKF SHAKF, and SHAFKF-P adaptation filters because of that the measurement noise covariance tunes by the adaptive transitive factor $a_2(\mathrm{k})$ to reduce the influence of uncertainty in measurement noise of the sensor.

8. Conclusions

In this chapter, the MEMS gyroscope drift is modeled by using ARMA (2, 1) for characterizing the MEMS gyro noise behavior. Moreover, ARMA-based linear Sage-Husa adaptive fading Kalman filter with double transitive factors is proposed. In the proposed algorithm, double adaptive transitive factors are used to update in the predicted state vector and measurement noise covariance matrix. The suggested algorithm is used to reduce the drift and random noise in the presence of MEMS gyroscope. From the AV analysis, the noise terms of ARW and Bs are reduced by order of 100. The proposed SHAFKF outperforms the CKF, IAE-AKF, and SHAKF algorithms in static case. It concludes that the SHAFKF algorithm is suitable for MEMS gyroscope signal drift minimization.

Author details

Mundla Narasimhappa
Electrical Engineering, Indian Institute of Technology Madras (IITM), Chennai, India

*Address all correspondence to: mr.narasimha08@gmail.com

References

[1] Martin P, Salaun E. Design and implementation of a low-cost observer-based attitude and heading reference system. Control Engineering Practice. 2010;**18**(7):712-722

[2] Li W, Wang J. Effective adaptive Kalman filter for MEMS-IMU/magnetometers integrated attitude and heading reference systems. The Journal of Navigation. 2013;**66**(1):99-113

[3] Gebre-Egziabher D, Hayward RC, Powel JD. A low-cost GPS/inertial attitude heading reference system (AHRS) for general aviation applications. In: Proceedings of the IEEE Symposium on Position Location and Navigation (PLANS). Palm Springs, CA, USA: IEEE; 1998. pp. 518-525

[4] Quinchia AG, Ferrer C, Falco G, Falletti E, Dovis F. Analysis and modelling of MEMS inertial measurement unit. In: Proceedings of the 2012 International Conference on Localization and GNSS; 2012. pp. 1-7

[5] Lee JK, Park EJ, Robinovitch SN. Estimation of attitude and external acceleration using inertial sensor measurement during various dynamic conditions. IEEE Transactions on Instrumentation and Measurement. 2012;**61**(8):2262-2273

[6] IEEE standard specification format guide and test procedure for single-axis interferometric fiber optic gyros. IEEE std 952-1997; 1998. pp. 1-84

[7] El-Diasty M, Pagiatakis S. A rigorous temperature dependent stochastic modelling and testing for MEMS-based inertial sensor errors. Sensors. 2009; **9**(11):8473-8489

[8] Chen WC, Gao GW, Wang J, Liu LL, Li XL. The study of the MEMS gyro zero drift signal based on the adaptive Kalman filter. Key Engineering Materials. 2012;**500**:635-639

[9] El-Rabbany A, El-Diasty M. An efficient neural network model for de-noising of MEMS-based inertial data. The Journal of Navigation. 2004;**57**(3): 407-415

[10] Wu X, Li Q. Research of the random noise compensation of MEMS gyro. In: System Simulation and Scientific Computing. Shanghai, China: Springer; 2012. pp. 328-335

[11] Kirkko-Jaakkola M, Collin J, Takala J. Bias prediction for MEMS gyroscopes. IEEE Sensors Journal. 2012; **12**(6):2157-2163

[12] Aggarwal P, Syed Z, Niu X, El-Sheimy N. A standard testing and calibration procedure for low cost MEMS inertial sensors and units. The Journal of Navigation. 2008;**61**(2): 323-336

[13] Yang G, Liu Y, Li M, Song S. AMA- and RWE-based adaptive Kalman filter for denoising fiber optic gyroscope drift signal. Sensors. 2015;**15**(10):26940-26960

[14] Huang L. Auto regressive moving average (ARMA) modeling method for gyro random noise using a robust Kalman filter. Sensors. 2015;**15**(10): 25277-25286

[15] Narasimhappa M, Rangababu P, Sabat SL, Nayak J. A modified Sage-Husa adaptive Kalman filter for de-noising fiber optic gyroscope signal. In: Proceedings of the 2012 Annual IEEE India Conference (INDICON); Kerala, India; 2012. pp. 1266-1271

[16] El-Sheimy N, Hou H, Niu X. Analysis and modeling of inertial sensors using Allan variance. IEEE

Transactions on Instrumentation and Measurement. 2008;**57**(1):140-149

[17] Sun J, Xu X, Liu Y, Zhang T, Li Y. FOG random drift signal de-noising based on the improved AR model and modified Sage-Husa adaptive Kalman filter. Sensors. 2016;**16**(7):1-19

[18] Kownacki C. Optimization approach to adapt Kalman filters for the real-time application of accelerometer and gyroscope signals' filtering. Digital Signal Processing. 2011;**21**(1):131-140

[19] Tanenhaus M, Carhoun D, Geis T, Wan E, Holland A. Miniature IMU/INS with optimally fused low drift MEMS gyro and accelerometers for applications in GPS-denied environments. In: Proceedings of the IEEE Symposium on 2012 IEEE/ION Position Location and Navigation Symposium (PLANS); IEEE; 2012. pp. 259-264

[20] Mohamed A, Schwarz K. Adaptive Kalman filtering for INS/GPS. Journal of Geodesy. 1999;**73**(4):193-203

[21] Grewal MS, Andrews AP. Kalman Filtering: Theory and Practice with MATLAB. Hoboken, New Jersey: John Wiley and Sons; 2015

[22] Hide C, Moore T, Smith M. Adaptive Kalman filtering for low-cost INS/GPS. The Journal of Navigation. 2003;**56**(1):143-152

[23] Yang Y, Xu T. An adaptive Kalman filter based on Sage windowing weights and variance components. The Journal of Navigation. 2003;**56**(02):231-240

[24] Yang Y, Gao W. Comparison of adaptive factors in Kalman filters on navigation results. Journal of Navigation. 2005;**58**(03):471-478

[25] Waegli A, Skaloud J, Guerrier S, Pares ME, Colomina I. Noise reduction and estimation in multiple micro-electromechanical inertial systems. Measurement Science and Technology. 2010;**21**(6):065201

[26] Moghaddamjoo A, Kirlin RL. Robust adaptive Kalman filtering with unknown inputs. IEEE Transactions on Acoustics, Speech, and Signal Processing. 1989;**37**(8):1166-1175

[27] Wang Y, Li N, Chen X, Liu M. Design and implementation of an AHRS based on MEMS sensors and complementary filtering. Advances in Mechanical Engineering. 2014;**6**:214726

[28] Narasimhappa M, Sabat SL, Nayak J. Fiber-optic gyroscope signal de-noising using an adaptive Robust Kalman filter. IEEE Sensors Journal. 2016;**16**(10):3711-3718

[29] Narasimhappa M, Mahindrakar AD, Guizilini VC, Terra MH, Sabat SL. An improved Sage Husa adaptive robust Kalman filter for de-noising the MEMS IMU drift signal. In: Proceedings of the IEEE Conference on Indian Control Conference (ICC), 2018. Kanpur, India: IEEE; 2018. pp. 229-234

[30] Narasimhappa M, Nayak J, Terra MH, Sabat SL. ARMA model based adaptive unscented fading filter for reducing drift of fiber optic gyroscope. Sensors and Actuators A: Physical. 2016;**251**:42-51

[31] Park M, Gao Y. Error and performance analysis of MEMS-based inertial sensors with a low-cost GPS receiver. Sensors. 2008;**8**(4):2240-2261

[32] Bistrov V. Performance analysis of alignment process of MEMS IMU. International Journal of Navigation and Observation. 2012;**2012**(731530):1-11

[33] Li Y, Hu B, Qin F, Li K. Online estimation of ARW coefficient of fiber optic gyro. Mathematical Problems in Engineering. 2014;**2014**(768590):1-10

[34] Georgy J, Noureldin A, Korenberg MJ, Bayoumi MM. Modeling the stochastic drift of a MEMS-based gyroscope in gyro/odometer/GPS integrated navigation. IEEE Transactions on Intelligent Transportation Systems. 2010;**11**(4): 856-872

[35] Almagbile A, Wang J, Ding W. Evaluating the performances of adaptive Kalman filter methods in GPS/ INS integration. Journal of Global Positioning Systems. 2010;**9**(1):33-40

[36] Simon D. Optimal State Estimation: Kalman, H Infinity, and Nonlinear Approaches. John Wiley and Sons; 2006

[37] MTi-G user manual and technical documentation. Revision H, Xsens Technologies B.V; 2010. pp. 1-64

Chapter 5

Discrete-Time Nonlinear Attitude Tracking Control of Spacecraft

Yuichi Ikeda

Abstract

Recent space programs require agile and large-angle attitude maneuvers for applications in various fields such as observational astronomy. To achieve agility and large-angle attitude maneuvers, it will be required to design an attitude control system that takes into account nonlinear motion because agile and large-angle rotational motion of a spacecraft in such missions represents a nonlinear system. Considerable research has been done about the nonlinear attitude tracking control of spacecraft, and these methods involve a continuous-time control framework. However, since a computer, which is a digital device, is employed as a spacecraft controller, the control method should have discrete-time control or sampled-data control framework. This chapter considers discrete-time nonlinear attitude tracking control problem of spacecraft. To this end, a Euler approximation system with respect to tracking error is first derived. Then, we design a discrete-time nonlinear attitude tracking controller so that the closed-loop system consisting of the Euler approximation system becomes input-to-state stable (ISS). Furthermore, the exact discrete-time system with a derived controller is indicated semiglobal practical asymptotic (SPA) stable. Finally, the effectiveness of proposed control method is verified by numerical simulations.

Keywords: spacecraft, attitude tracking control, discrete-time nonlinear control

1. Introduction

Recent space programs require agile and large-angle attitude maneuvers for applications in various fields such as observational astronomy [1–3]. To achieve agility and large-angle attitude maneuvers, it will be required to design an attitude control system that takes into account nonlinear motion because agile and large-angle rotational motion of a spacecraft in such missions represents a nonlinear system.

Considerable research has been done about the nonlinear attitude tracking control of spacecraft [4–12], and these methods involve a continuous-time control framework. However, since a computer, which is a digital device, is employed as a spacecraft controller, the control method should have discrete-time control or sampled-data control framework.

Although a sampled-data control method for nonlinear system did not advance because it is difficult to discretize a nonlinear system, a control method based on the Euler approximate model has been proposed in recent years [13, 14] and is applied to ship control [15]. Although our research group has proposed a sampled-data

IntechOpen

control method using backstepping [16] and a discrete-time control method based on sliding mode control [17] for spacecraft control problem, these methods are disadvantageous because control input amplitude depends on the sampling period T as the control law is of the form $u = a(x) + (b(x)/T)$.

For these facts, about the spacecraft attitude control problem that requires agile and large-angle attitude maneuvers, this chapter proposed a discrete-time nonlinear attitude tracking control in which the control input amplitude is independent of the sampling period T. The effectiveness of proposed control method is verified by numerical simulations.

The following notations are used throughout the chapter. Let R and N denote the real and the integer numbers. Rn and $R^{n \times m}$ are the sets of real vectors and matrices. For real vector $a \in R^n$, a^T is the vector transpose, $\|a\|$ denotes the Euclidean norm, and $a^\times \in R^{3\times3}$ is the skew symmetric matrix

$$a^\times = \begin{bmatrix} 0 & -a_3 & a_2 \\ a_3 & 0 & -a_1 \\ -a_2 & a_1 & 0 \end{bmatrix}$$

derived from vector $a \in R^3$. For real symmetric matrix A, $A > 0$ means the positive definite matrix. The identity matrix of size 3×3 is denoted by I_3. $\lambda_A^{\max} \in R$ and $\lambda_A^{\min} \in R$ are the maximal and the minimal eigenvalues of a matrix A, respectively.

2. Relative equation of motion and discrete-time model for spacecraft

In this chapter, as the kinematics represents the attitude of the spacecraft with respect to the inertia frame $\{i\}$, the modified Rodrigues parameters (MRPs) [5] are used. The rotational motion equations of the spacecraft's body-fixed frame $\{b\}$ are given by the following equations:

$$\dot{\sigma}(t) = G(\sigma(t))\omega(t), \tag{1}$$

$$G(\sigma(t)) = \frac{1}{2}\left\{\frac{1-\|\sigma(t)\|^2}{2} I_3 + \sigma(t)\sigma(t)^T + \sigma(t)^\times\right\},$$

$$\dot{\omega}(t) = J^{-1}\{-\omega(t)^\times J\omega(t) + u(t) + w(t)\}, \tag{2}$$

where Eq. (1) is the kinematics that represents the attitude of $\{b\}$ with respect to the $\{i\}$, Eq. (2) is the rotation dynamics, $\sigma(t) \in R^3$ [−] is the MRPs, $\omega(t) \in R^3$ [rad/s] is the angular velocity, $u(t) \in R^3$ [Nm] is the control torque (input), $w(t) \in R^3$ [Nm] is the disturbance input, and $J \in R^{3\times3}$ [kg m^2] is the moment of inertia.

We consider a control problem in which a spacecraft tracks a desired attitude (MRPs) $\sigma_d(t) \in R^3$ and angular velocity $\omega_d(t) \in R^3$ in fixed frame $\{d\}$. The MRPs of the relative attitude $\sigma_e(t) \in R^3$ and the relative angular velocity $\omega_e(t) \in R^3$ in the frame $\{b\}$ are given by

$$\sigma_e(t) = \frac{N_e(t)}{1 + \|\sigma(t)\|^2\|\sigma_d(t)\|^2 + 2\sigma_d(t)^T\sigma(t)}, \tag{3}$$

$$N_e(t) = \left(1 - \|\sigma_d(t)\|^2\right)\sigma(t) - \left(1 - \|\sigma(t)\|^2\right)\sigma_d(t) + 2\sigma(t)^\times\sigma_d(t),$$

$$\omega_e(t) = \omega(t) - C(t)\omega_d(t), \tag{4}$$

where $C(t) \in R^{3\times3}$ is the direction cosine matrix from $\{b\}$ to $\{d\}$ that expresses the following Eq. [7]:

$$C(t) = I_3 + \frac{8(\sigma_e(t)^\times)^2 - 4\left(1 - \|\sigma_e(t)\|^2\right)\sigma_e(t)^\times}{\left(1 + \|\sigma_e(t)\|^2\right)^2}. \tag{5}$$

Substituting Eqs. (3) and (4) into Eqs. (1) and (2) using the identity $\dot{C}(t) = -\omega_e(t)^\times C(t)$ yields the following relative motion equations:

$$\dot{\sigma}_e(t) = G(\sigma_e(t))\omega_e(t), \tag{6}$$

$$\begin{aligned}\dot{\omega}_e(t) = & J^{-1}[-\{\omega_e(t) + C(t)\omega_d(t)\}^\times J\{\omega_e(t) + C(t)\omega_d(t)\} \\ & -J\{C(t)\dot{\omega}_d(t) - \omega_e(t)^\times C(t)\omega_d(t)\} + u(t) + w(t)]\end{aligned} \tag{7}$$

Hereafter, we assume that the variables of spacecraft $\sigma(t)$ and $\omega(t)$ are directly measurable and J is known. In addition, regarding the desired states $\sigma_d(t)$, $\omega_d(t)$, $\dot{\omega}_d(t)$, and the disturbance $w(t)$, the following assumption is made.

Assumption 1: the desired states $\sigma_d(t)$, $\omega_d(t)$, and $\dot{\omega}_d(t)$ are uniformly continuous and bounded $\forall t \in [0, \infty)$. The disturbance $w(t)$ is uniformly bounded $\forall t \in [0, \infty)$.

From Eqs. (A4) and (A5) in Appendix, the exact discrete-time model of relative motion equations is obtained as

$$\sigma_{e,k+1} = \sigma_{e,k} + \int_{kT}^{(k+1)T} G(\sigma_e(s))\omega_e(s)\, ds, \tag{8}$$

$$\begin{aligned}\omega_{e,k+1} = \omega_{e,k} + \int_{kT}^{(k+1)T} & [-\{\omega_e(s) + C(s)\omega_d(s)\}^\times J\{\omega_e(s) + C(s)\omega_d(s)\} \\ & -J\{C(s)\dot{\omega}_d(s) - \omega_e(s)^\times C(s)\omega_d(s)\} + u_k + w_k]ds\end{aligned} \tag{9}$$

and the Euler approximate model of relative motion equations are obtained as

$$\sigma_{e,k+1} = \sigma_{e,k} + TG(\sigma_{e,k})\omega_{e,k}, \tag{10}$$

$$\begin{aligned}\omega_{e,k+1} = \omega_{e,k} - TJ^{-1}[& -\{\omega_{e,k} + C_k\omega_{d,k}\}^\times J\{\omega_{e,k} + C_k\omega_{d,k}\} \\ & -J\{C_k\dot{\omega}_{d,k} - \omega_{e,k}{}^\times C_k\omega_{d,k}\} + u_k + w_k].\end{aligned} \tag{11}$$

3. Discrete-time nonlinear attitude tracking control

We derive a controller based on the backstepping approach that makes the closed-loop system consisting of the Euler approximate modes (10) and (11) become input-to-state stable (ISS), i.e., the state variable of closed-loop system $x_k = \left[\sigma_{e,k}^T \omega_{e,k}^T\right]^T$ satisfies the following equation:

$$\|x_{k+1}\| \le \rho(\|x_0\|, k) + \gamma(\|w_k\|), \quad \forall x_k \in R^3, \quad \forall w_k \in R^3,$$

where $\rho(\cdot)$ is the class KL function and $\gamma(\cdot)$ is the class K function. To this end, assume that $\omega_{e,k}$ is the virtual input to subsystem (10), and derive the stabilizing function α_k that $\sigma_{e,k}$ is asymptotic convergence to zero. Then, derive the control input u_k that closed-loop system becomes ISS. Here, regarding the variable $\sigma_{e,k}$, the following assumption is made.

Assumption 2: $\sigma_{e,k}$ lies in the region that satisfies the following equation:

$$0 \leq \|\sigma_{e,k}\| \leq 1, \quad \forall k.$$

Remark 1: from the relational expression

$$\sigma_{e,k} = \frac{\varepsilon_{e,k}}{1+\eta_{e,k}},$$

where $\varepsilon_{e,k} \in \mathrm{R}^3$ and $\eta_{e,k} \in \mathrm{R}$ are the quaternion $\left(\left\| \left[\varepsilon_{e,k}^T \eta_{e,k} \right]^T \right\| = 1, \|\varepsilon_{e,k}\| \leq 1, |\eta_{e,k}| \leq 1, \forall k \right)$. Assumption 2 is equivalent to $\eta_{e,k} \in [0, 1]$.

In addition, Lemmas when using the derivation of the control law are shown below.

Lemma 1: for all $\sigma \in \mathrm{R}^3$, the following equations hold [5]:

$$\sigma^T G(\sigma) = b\sigma^T, \quad G(\sigma)^T G(\sigma) = b^2 I_3, \quad \left(b = \frac{1+\|\sigma\|^2}{4} > 0 \right).$$

Lemma 2: when the quadratic equation

$$ax^2 + bx + c = 0 (a, b, c \in \mathrm{R})$$

has two distinct real roots $x = \alpha, \beta (\alpha < \beta)$, if $a > 0$, then the solution of the quadratic inequality

$$ax^2 + bx + c < 0$$

is $\alpha < x < \beta$.

3.1 Derivation of virtual input α_k

Assume that $\omega_{e,k}$ is the virtual input to subsystem (10), and define the stabilizing function such that

$$\omega_{e,k} = \alpha_k = -f_1 \sigma_{e,k}, \tag{12}$$

where $f_1 \in \mathrm{R}$ is the feedback gain. The candidate Lyapunov function for (10) is defined as

$$V_1(k) = \|\sigma_{e,k}\|^2. \tag{13}$$

From Lemma 1, the difference of Eq. (13) along the trajectories of the closed-loop system is given by

$$\Delta V_1(k) = V_1(k+1) - V_1(k) = \left\{ (Tf_1 b_k)^2 - 2Tf_1 b_k \right\} \|\sigma_{e,k}\|^2. \tag{14}$$

From Lemma 2, $\Delta V_1(k)$ becomes negative, i.e., the range of f_1 that holds the following equation

$$(Tf_1 b_k)^2 - 2Tf_1 b_k < 0 \tag{15}$$

is obtained as

$$0 < f_1 < \frac{2}{Tb_k}. \tag{16}$$

In addition, since $2 \leq (1/b_k) \leq 4$ under Assumption 2, the range of f_1 that holds Eq. (15) is obtained as

$$0 < f_1 < \frac{4}{T}. \tag{17}$$

Therefore, if f_1 satisfies Eq. (17) and $\omega_{e,k} \to \alpha_k (k \to \infty)$, then $\sigma_{e,k} \to 0$.

3.2 Derivation of control input u_k

The error variable between the state $\omega_{e,k}$ and α_k is defined as

$$z_k := \omega_{e,k} - \alpha_k. \tag{18}$$

The control input u_k that makes the closed-loop system becomes ISS is derived. From Eq. (18), subsystem (10) becomes

$$\sigma_{e,k+1} = \sigma_{e,k} + TG(\sigma_{e,k})(z_k + \alpha_k). \tag{19}$$

From Eqs. (18) and (19) and the following equation

$$\alpha_k - \alpha_{k+1} = Tf_1\{G(\sigma_{e,k})z_k - f_1 b_k \sigma_{e,k}\},$$

the discrete-time equation with respect to z_k is

$$\begin{aligned} z_{k+1} = {} & z_k + Tf_1\{G(\sigma_{e,k})z_k - f_1 b_k \sigma_{e,k}\} \\ & + TJ^{-1}[-\{z_k + \alpha_{,k} + C_k \omega_{d,k}\} \times J\{z_k + \alpha_{,k} + C_k \omega_{d,k}\} \\ & - J\{C_k \dot{\omega}_{d,k} - (z_k + \alpha_{,k}) \times C_k \omega_{d,k}\} + u_k + w_k]. \end{aligned} \tag{20}$$

Now, by setting u_k to

$$\begin{aligned} u_k = {} & \{z_k + \alpha_{,k} + C_k \omega_{d,k}\} \times J\{z_k + \alpha_{,k} + C_k \omega_{d,k}\} \\ & + J\{C_k \dot{\omega}_{d,k} - (z_k + \alpha_{,k}) \times C_k \omega_{d,k}\} \\ & - f_1 J\{G(\sigma_{e,k})z_k - f_1 b_k \sigma_{e,k}\} - f_2 J z_k, \end{aligned}$$

Eq. (20) becomes

$$z_{k+1} = (1 - Tf_2)z_k + TJ^{-1}w_k, \tag{21}$$

where $f_2 \in \mathrm{R}$ is the feedback gain. The candidate Lyapunov function for Eqs. (19) and (21) is defined as

$$V_2(k) = V_1(k) + \|z_k\|^2 = \|X_k\|^2, X_k = \left[\sigma_{e,k}^T z_k^T\right]^T. \tag{22}$$

As Eq. (14) is given by

$$\Delta V_1(k) = (Tb_k)^2\|z_k\|^2 + \left\{(Tf_1 b_k)^2 - 2Tf_1 b_k\right\}\|\sigma_{e,k}\|^2 + 2Tb_k(1 - Tf_1 b_k)z_k^T\sigma_{e,k}^T$$

from Eq. (18), by using completing square, the difference of Eq. (22) along the trajectories of the closed-loop system is given by

$$\begin{aligned}\Delta V_2(k) &= (T^2 f_2^2 - 2Tf_2 + T^2 b_k^2)\|z_k\|^2 + \left\{(Tf_1 b_k)^2 - 2Tf_1 b_k\right\}\|\sigma_{e,k}\|^2 \\ &\quad + 2Tb_k(1 - Tf_1 b_k)z_k^T\sigma_{e,k}^T + T^2 w_k^T J^{-2} w_k + 2T(1 - Tf_2)w_k^T J^{-1} z_k \\ &\le (2T^2 f_2^2 - 4Tf_2 + T^2 b_k^2 + 1)\|z_k\|^2 + \left\{(Tf_1 b_k)^2 - 2Tf_1 b_k\right\}\|\sigma_{e,k}\|^2 \\ &\quad + 2Tb_k(1 - Tf_1 b_k)z_k^T\sigma_{e,k}^T + 2\left(\frac{T}{\lambda_J}\right)^2\|w_k\|^2 \\ &= X_k^T Q_k X_k + 2\left(\frac{T}{\lambda_J}\right)^2\|w_k\|^2,\end{aligned} \tag{23}$$

where

$$\lambda_J = \|J\|,\ Q_k = \begin{bmatrix} Q_{11,k} & Q_{12,k} \\ Q_{12,k}^T & Q_{22,k}\end{bmatrix},\ Q_{11,k} = \left\{(Tf_1 b_k)^2 - 2Tf_1 b_k\right\}I_3,$$

$$Q_{12,k} = Tb_k(1 - Tf_1 b_k)I_3,\ Q_{22,k} = (2T^2 f_2^2 - 4Tf_2 + T^2 b_k^2 + 1)I_3.$$

In Eq. (23), if $Q_k < 0$, then

$$\Delta V_2(k) \le -\left|\lambda_{Q_k}^{\min}\right|\|X_k\|^2 + 2\left(\frac{T}{\lambda_J}\right)^2\|w_k\|^2,$$

where $\lambda_{Q_k}^{\min} < 0 \in R$ is the minimum eigenvalue of Q_k and the condition of ISS holds [18]. Hereafter, conditions of f_1 and f_2 which the matrix Q_k holds $Q_k < 0$ are derived under Assumption 2.

From Schur complement, condition $Q_k < 0$ is equivalent to the following equations:

$$(Tf_1 b_k)^2 - 2Tf_1 b_k < 0, \tag{24}$$

$$2T^2 f_2^2 - 4Tf_2 + c_k < 0 \quad \left(c_k = \frac{Tb_k f_1^2 - 2f_1 - Tb_k}{Tb_k f_1^2 - 2f_1}\right). \tag{25}$$

Condition (24) is the same as Eq. (15), and assume that Eq. (24) holds. From Lemma 2, the range of f_2 that holds for Eq. (25) is obtained as

$$\frac{2 - \sqrt{2(2 - c_k)}}{2T} < f_2 < \frac{2 + \sqrt{2(2 - c_k)}}{2T}, \tag{26}$$

and the following Eq.

$$2 - c_k > 0 \quad \Rightarrow \quad \frac{Tb_k f_1^2 - 2f_1 + Tb_k}{Tb_k f_1^2 - 2f_1} > 0 \tag{27}$$

must hold true in order to obtain a real number. As the denominator of Eq. (27) is the same as Eq. (24), the following equation must hold

$$Tb_k f_1^2 - 2f_1 + Tb_k < 0 \tag{28}$$

in order to hold Eq. (27). From Lemma 2, the range of f_1 that holds for Eq. (28) is obtained as

$$\frac{1-\sqrt{1-(Tb_k)^2}}{Tb_k} < f_1 < \frac{1+\sqrt{1-(Tb_k)^2}}{Tb_k}, \tag{29}$$

and the following Eq.

$$1-(Tb_k)^2 > 0 \quad \Rightarrow \quad 0 < T < \frac{1}{b_k} \tag{30}$$

must hold in order to have the real number. As $2 \le (1/b_k) \le 4$ under Assumption 2, T must satisfy the condition

$$0 < T < 2. \tag{31}$$

In addition, since

$$\max_{b_k} \frac{1-\sqrt{1-(Tb_k)^2}}{Tb_k} = \frac{2-\sqrt{4-T^2}}{T},$$

$$\min_{b_k} \frac{1+\sqrt{1-(Tb_k)^2}}{Tb_k} = \frac{2+\sqrt{4-T^2}}{T}$$

under Assumption 2, the condition (29) is given by

$$\frac{2-\sqrt{4-T^2}}{T} < f_1 < \frac{2+\sqrt{4-T^2}}{T} (0 < T < 2). \tag{32}$$

Therefore, if f_1 satisfies Eq. (32) under Assumption 2, Eqs. (27) and (28) hold. Furthermore, since

$$\max_{b_k} \frac{2-\sqrt{2(2-c_k)}}{2T} = \frac{1}{T} - \sqrt{\frac{Tf_1^2 - 4f_1 + T}{2T^2 f_1 (Tf_1 - 4)}},$$

$$\min_{b_k} \frac{2+\sqrt{2(2-c_k)}}{2T} = \frac{1}{T} + \sqrt{\frac{Tf_1^2 - 4f_1 + T}{2T^2 f_1 (Tf_1 - 4)}},$$

under Assumption 2, the condition (26) is given by.

$$\frac{1}{T} - \sqrt{\frac{Tf_1^2 - 4f_1 + T}{2T^2 f_1 (Tf_1 - 4)}} < f_2 < \frac{1}{T} + \sqrt{\frac{Tf_1^2 - 4f_1 + T}{2T^2 f_1 (Tf_1 - 4)}} (0 < T < 2). \tag{33}$$

Therefore, if f_1 and f_2 satisfy Eqs. (32) and (33) under Assumption 2, then $Q_k < 0$.

Summarizing the above, the following theorem can be obtained.

Theorem 1: if sampling period T and feedback gains f_1 and f_2 satisfy Eqs. (31), (32), and (33) under Assumption 2, then the closed-loop systems (10) and (11) with the following control law

$$\begin{aligned} u_k &= \{z_k + \alpha_{,k} + C_k\omega_{d,k}\}^{\times} J\{z_k + \alpha_{,k} + C_k\omega_{d,k}\} \\ &\quad + J\{C_k\dot{\omega}_{d,k} - (z_k + \alpha_{,k})^{\times} C_k\omega_{d,k}\} \\ &\quad - f_1 J\{G(\sigma_{e,k})z_k - f_1 b_k \sigma_{e,k}\} - f_2 J z_k \\ &= \omega_k{}^{\times} J\omega_k + J\{C_k\dot{\omega}_{d,k} - (z_k + \alpha_{,k})^{\times} C_k\omega_{d,k}\} \\ &\quad - f_1 f_2 J\sigma_{e,k} - J\{f_1 G(\sigma_{e,k}) + f_2 I_3\}\omega_{e,k} \end{aligned} \tag{34}$$

becomes ISS.

Then, we show that the pair $(u_k, V_2(k))$ is semiglobal practical asymptotic (SPA) stabilizing pair for the Euler approximate systems (10) and (11). Hereafter, suppose that sampling period T and feedback gains f_1 and f_2 satisfy Eqs. (31), (32), and (33) under Assumption 2. By using the following coordinate transformation

$$X_k = \begin{bmatrix} 1 & 0 \\ f_1 & 1 \end{bmatrix} \begin{bmatrix} \sigma_{e,k} \\ \omega_{e,k} \end{bmatrix} = Z\overline{X}_k,$$

Lyapunov function $V_2(k)$ and its difference $\Delta V_2(k)$ can be rewritten as

$$V_2(k) = \overline{X}_k^T Z^T Z \overline{X}_k = \overline{X}_k^T R \overline{X}_k,$$

$$\Delta V_2(k) = \overline{X}_k^T Z^T Q_k Z \overline{X}_k + 2\left(\frac{T}{\lambda_J}\right)^2 \|w_k\|^2 = \overline{X}_k^T \overline{Q}_k \overline{X}_k + 2\left(\frac{T}{\lambda_J}\right)^2 \|w_k\|^2.$$

Since $R > 0$ and $\overline{Q}_k < 0$, $V_2(k)$ and $\Delta V_2(k)$ satisfy following equations:

$$\lambda_R^{\min} \|\overline{X}_k\|^2 \le V_2(k) \le \lambda_R^{\max} \|\overline{X}_k\|^2, \tag{35}$$

$$\Delta V_2(k) \le -\left|\lambda_{\overline{Q}_k}^{\min}\right| \|\overline{X}_k\|^2 + 2\left(\frac{T}{\lambda_J}\right)^2 \|w_k\|^2. \tag{36}$$

In addition, $\overline{X}_k$ is bounded, and $V_2(k)$ is radially unbounded from Eqs. (35) and (36). Hence, the control input (34) satisfies the following equation under Assumption 1:

$$\|u_k\| \le M, \tag{37}$$

where M is a positive constant. Furthermore, $V_2(k)$ also satisfies the following equation for all $x, z \in \mathrm{R}^6$ with $\max\{\|x\|, \|z\|\} \le \Delta$:

$$\begin{aligned} |V_2(x) - V_2(z)| &= \left|x^T R x - z^T R z\right| = \left|(x+z)^T R (x - z)\right| \\ &= \lambda_R^{\max} \|x + z\| \|x - z\| \le 2\Delta \lambda_R^{\max} \|x - z\|, \end{aligned} \tag{38}$$

where Δ is a positive constant. Therefore, from Eqs. (35) to (38), Lyapunov function $V_2(k)$ and control input u_k satisfied Eqs. (A8)–(A11) in Definition 2 under Assumptions 1 and 2, and the pair $(u_k, V_2(k))$ becomes SPA stabilizing pair for the

Euler approximate systems (10) and (11). Then, the following theorem can be obtained by Theorem A.1 in Appendix.

Theorem 2: control input (34) is SPA stabilizing for exact discrete-time systems (8) and (9).

4. Numerical simulation

The properties of the proposed method are discussed in the numerical study. For this purpose, parameter setting of simulation is as follows:

$$J = \begin{bmatrix} 7050.0 & -0.536 & 43.9 \\ -0.536 & 2390 & 1640.0 \\ 43.9 & 1640.0 & 6130.0 \end{bmatrix} \text{kgm}^2, \sigma(0) = \begin{bmatrix} 0 \\ 0 \\ 0 \end{bmatrix}, \omega(0) = \begin{bmatrix} 0 \\ 0 \\ 0 \end{bmatrix} \text{ rad/s}$$

$$T = \begin{cases} 1.0 : \text{Case 1} \\ 0.5 : \text{Case 2}, f_1 = 0.6, f_2 = 0.8. \\ 0.1 : \text{Case 3} \end{cases}$$

The moment of inertia J is from [1]. The initial values $\sigma(0)$ correspond to Euler angles of 1–2–3 system of $\theta(0) = [\theta_1(0)\theta_2(0)\theta_3(0)]^T = [0\ \ 0\ \ 0]^T$ [deg]. The feedback gains f_1 and f_2 satisfy Eqs. (25) and (28) for all cases of T. The desired states $\sigma_d(t)$, $\omega_d(t)$, and $\dot{\omega}_d(t)$ in this simulation are the switching maneuver as shown in **Figure 1**.

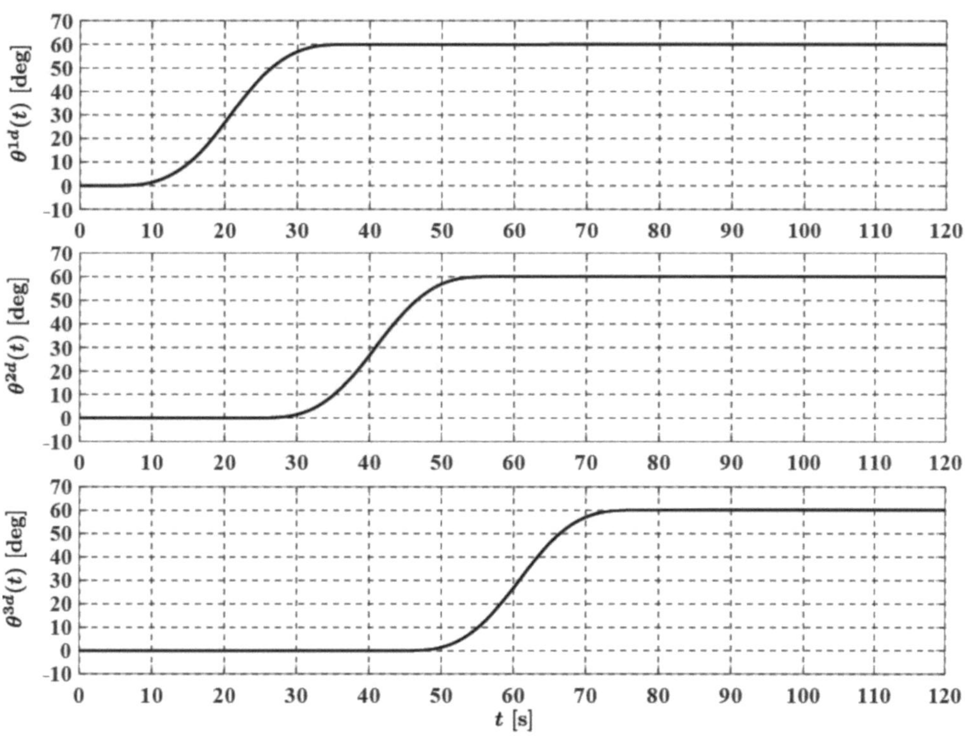

Figure 1.
Switching maneuver.

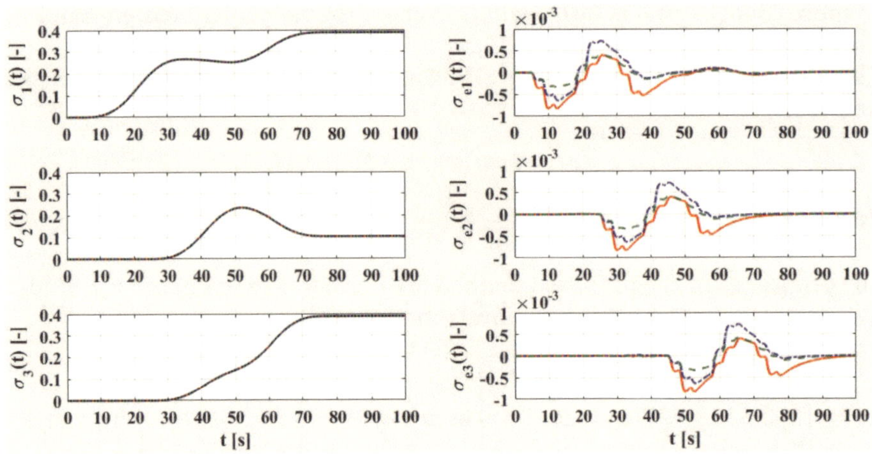

Figure 2.
Time histories of MRPs $\sigma(t)$ and $\sigma_e(t)$ (solid line, case 1; dashed-dotted line, case 2; dashed line, and case 3; dotted line, $\sigma_d(t)$).

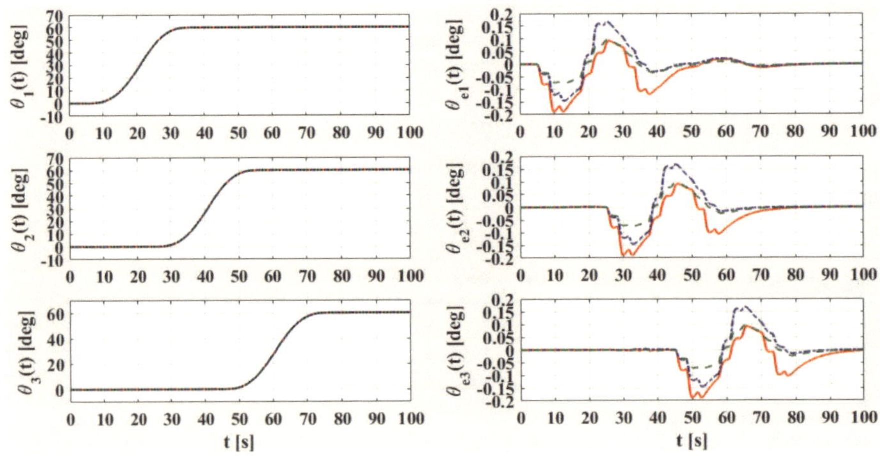

Figure 3.
Time histories of attitude angles $\theta(t)$ and $\theta_e(t)$ (solid line, case 1; dashed-dotted line, case 2; dashed line, and case 3; dotted line, $\theta_d(t)$).

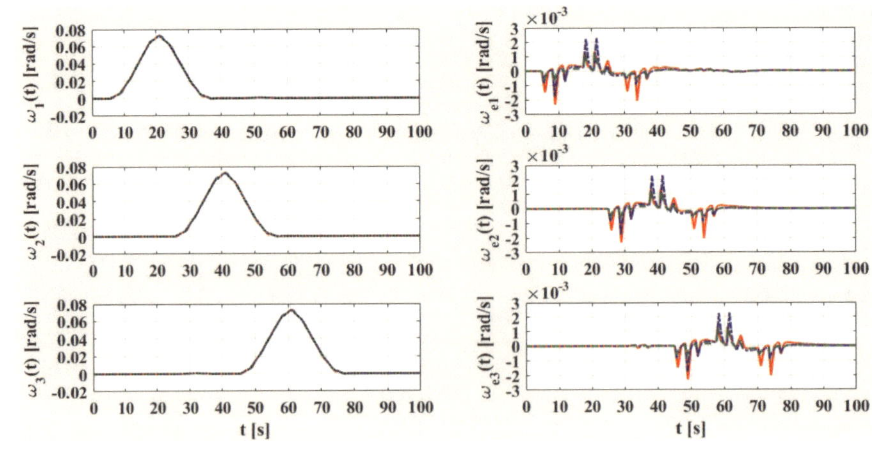

Figure 4.
Time histories of angular velocities $\omega(t)$ and $\omega_e(t)$ (solid line, case 1; dashed-dotted line, case 2; dashed line, and case 3; dotted line, $\omega_d(t)$).

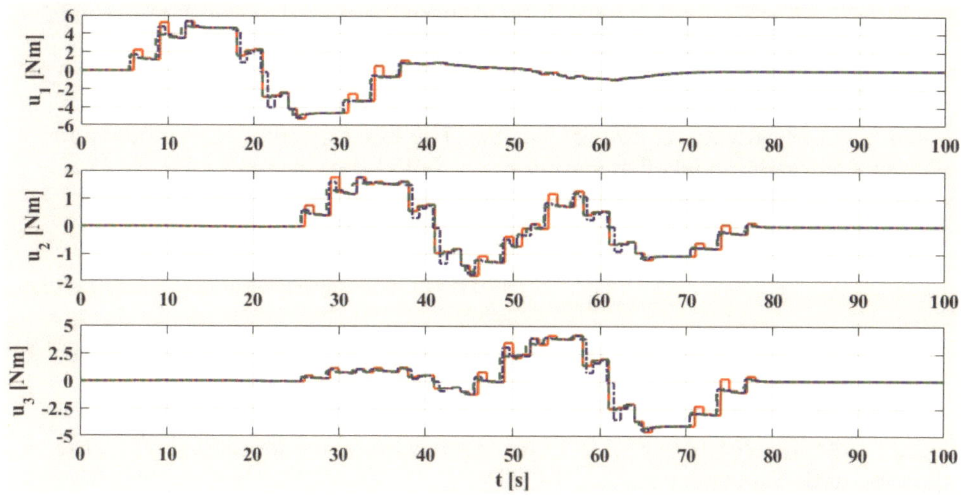

Figure 5.
Time histories of control input $u(t)$ (solid line, case 1; dashed-dotted line, case 2; and dashed line, case 3).

The results of the numerical simulation are shown in **Figures 2–5**. The relative attitude $\sigma_e(t)$ and relative angular velocity $\omega_e(t)$ converge to the neighborhood of $(\sigma_e(t), \omega_e(t)) = (0, 0)$, and the control input amplitude $u(t)$ does not depend on the sampling period T although there is a slight difference in the maximal value of $u(t)$.

5. Conclusion

This chapter considers the spacecraft attitude tracking control problem that requires agile and large-angle attitude maneuvers and proposed a discrete-time nonlinear attitude tracking control that the amplitude of the control input does not depend on the sampling period T. The effectiveness of proposed control method is verified by numerical simulations. Extension to the guarantee of stability as sampled-data control system will be subject to future work.

Appendix: sampled-data control of nonlinear system

This section shows preliminary results for nonlinear sampled-data control [13, 14, 19].

Let us consider the following nonlinear system:

$$\dot{x}(t) = f(x(t), u(t)), x(0) = x_0, f(0, 0) = 0, \tag{A1}$$

where $x(t) \in \mathrm{R}^n$ is the state variable and $x(t) \in \mathrm{R}^m$ is the control input. The function $f(x(t), u(t))$ in Eq. (A1) is assumed to be such that, for each initial condition and each constant control input, there exists a unique solution defined on some intervals of $x[0, \tau)$.

The nonlinear system (A1) is assumed to be between a sampler (A/D converter) and zero-order hold (D/A converter), and the control signal is assumed to be piecewise constant, that is,

$$u(t) = u(kT) =: u(k), \forall t \in [kT, (k+1)T], k \in \{0\} \cup \mathrm{N}, \tag{A2}$$

where $T > 0$ is a sampling period. In addition, assume that the state variable

$$x(k) := x(kT) \tag{A3}$$

is measurable at each sampling instance. The exact discrete-time model and Euler approximate model of the nonlinear sampled-data systems (A1)–(A3) are expressed as follows, respectively:

$$x_{k+1} = x_k + \int_{kT}^{(k+1)T} f(x(s), u_k)\, ds =: F_T^e(x_k, u_k), \tag{A4}$$

$$x_{k+1} = x_k + Tf(x_k, u_k) =: F_T^{Euler}(x_k, u_k), \tag{A5}$$

where we abbreviate $x(k)$ and $u(k)$ to x_k and u_k. For the stability of the exact discrete-time model (A4) (F_T^e) and Euler approximate model (A5) (F_T^{Euler}), the following definitions are used [13, 14, 19].

Definition 1: consider the following discrete-time nonlinear system:

$$x_{k+1} = F_T(x_k, u_T(x_k)), \tag{A6}$$

where $x_k \in R^n$ is the state variable and $u_T(x_k) \in R^m$ is a control input. The family of controllers $u_T(x_k)$ SPA stabilizes the system (A6) if there exists a class KL function $\beta(\cdot)$ such that for any strictly positive real numbers (D, ν), there exists $T^* > 0$, and such that for all $T \in (0, T^*)$ and all initial state x_0 with $\|x_0\| \le D$, the solution of the system satisfies

$$\|x_k\| \le \beta(\|x_0\|, kT) + \nu, \forall k \in \{0\} \cup N. \tag{A7}$$

Definition 2: let $\hat{T} > 0$ be given, and for each $T \in (0, \hat{T})$, let functions $V_T : R^n \to R$ and $u_T : R^n \to R^m$ be defined. The pair of families (u_T, V_T) is a SPA stabilizing pair for the system (A7) if there exist a class K_∞ functions α_1, α_2, and α_3 such that for any pair of strictly positive real numbers (Δ, δ), there exists a triple of strictly positive real numbers $(T^*, L, M)(T^* \le \hat{T})$ such that for all $x, z \in R^n$ with $\max\{\|x\|, \|z\|\} \le \Delta$, and $T \in (0, T^*)$:

$$\alpha_1(\|x\|) \le V_T(x) \le \alpha_2(\|x\|), \tag{A8}$$

$$V_T(F_T(x, u_T(x))) - V_T(x) \le -\alpha_3(\|x\|) + T\delta, \tag{A9}$$

$$|V_T(x) - V_T(z)| \le L\|x - z\|, \tag{A10}$$

$$\|u_T(x)\| \le M. \tag{A11}$$

In addition, if there exists $T^{**} > 0$ such that Eqs. (A8)–(A11) with $\delta = 0$ hold for all $x \in R^n$ and $T \in (0, T^{**})$, then the pair (u_T, V_T) is globally asymptotic (GA) stabilizing pair for the system (A6).

Using the above definitions, the following theorem is obtained by literatures [13, 14, 19].

Theorem A.1: if the pair (u_T, V_T) is SPA stabilizing for F_T^{Euler}, then u_T is SPA stabilizing for F_T^e.

Hence, if we can find a family of pairs of (u_T, V_T) that is a GA or SPA stabilizing pair for F_T^{Euler}, then the controller u_T will stabilize the exact model F_T^e.

Author details

Yuichi Ikeda
Shonan Institute of Technology, Fujisawa, Japan

*Address all correspondence to: ikeda@mech.shonan-it.ac.jp

References

[1] Kamiya T, Maeda K, Ogura N. Preshaping profile for flexible spacecraft rest-to-rest maneuver. In: AIAA Guidance, Navigation, and Control Conference and Exhibit; AIAA 2006–6181; 2006

[2] Somov Y, Butyrin S, Somov S. Guidance and robust gyromoment precise attitude control of agile observation spacecraft. In: 17th IFAC World Congress; 2008. pp. 3422-3427

[3] Nagashio T, Kida T, Mitani S, Ohtani T, Yamaguchi I, Kasai T, Hamada H. A preliminary controller design study on precise switching maneuver of flexible spacecraft astro-G. In: Proceedings of 25th Guidance Control Symposium (in Japanese); 2008. pp. 25-32

[4] Dalsmo M, Egeland O. State feedback H_∞-suboptimal control of a rigid spacecraft. IEEE Transactions on Automatic Control. 1997;**42**(8): 1186-1189

[5] Tsiotras P. Further passivity results for the attitude control problem. IEEE Transactions on Automatic Control. 1998;**43**(11):1597-1160

[6] DeVon DA, Fuentes RJ, Fausz JL. Passivity-based attitude control for an integrated power and attitude control system using variable speed control moment gyroscopes. In: Proceedings of the 2004 American Control Conference; 2004. pp. 1019-1024

[7] Meng Z, Ren W, You Z. Decentralized cooperative attitude tracking using modified rodriguez parameters. In: Proceedings of the 48th IEEE Conference on Decision and Control Held Jointly with 28th Chinese Control Conference; 2009. pp. 853-858

[8] Schlanbusch R, Loria A, Kristiansen R, Nicklasson P. J. PD+ attitude control of rigid bodies with improved performance. In: Proceedings of the 49th IEEE Conference on Decision and Control; 2010. pp. 7069-7074

[9] Liu S, Sun J, Geng Z. Passivity-based finite-time attitude control problem. In: 9th Asian Control Conference; 2013. pp. 1-6

[10] Cong BL, Chen Z, Liu XD. Robust attitude control with improved transient performance. In: 19th IFAC World Congress; 2014. pp. 463-468

[11] Ikeda Y, Kida T, Nagashio T. Nonlinear tracking control of rigid spacecraft under disturbance using PID type H_∞ adaptive state feedback. Transactions of the Japan Society for Aeronautical and Space Sciences. 2015; **58**(5):289-297

[12] Ikeda Y, Kida T, Nagashio T. Stabilizing nonlinear adaptive control of spacecraft before and after capture. Transactions of the Japan Society for Aeronautical and Space Sciences. 2016; **59**(1):1-9

[13] Nesi'c D, Teel AR, Kokotovi'c PV. Sufficient conditions for stabilization of sampled-data nonlinear systems via discrete-time approximation. Systems & Control Letters. 1999;**38**(4–5): 259-270

[14] Nesi'c D, Teel AR. Stabilization of sampled-data nonlinear system via backstepping on their Euler approximate model. Automatica. 2006; **42**(10):1801-1808

[15] Katayama H. Nonlinear sampled-data stabilization of dynamically positioned ships. IEEE Transactions on Control Systems Technology. 2010; **18**(2):463-468

[16] Nishimura M, Nagahio T, Kida T. Discrete-time tracking control system

design based on a practical stability for spacecraft. In: Proceedings of 53rd the Japan Joint Automatic Control Conference (in Japanese); 2010. pp. 895-900

[17] Ikeda Y. Nonlinear sampled-data control of spacecraft by sliding mode control. In: Proceedings of the 27th Guidance and Control Symposium (in Japanese); 2010. pp. 97-100

[18] Jiang ZP, Wang Y. Input-to-state stability for discrete-time nonlinear system. Automatica. 2001;**35**:857-869

[19] Laila DS, Nesi'c D, Astolfi A. Sampled-Data Control of Nonlinear Systems, Advanced Topics in Control Systems Theory. In: Lecture Notes in Control and Information Sciences. Springer; 2005. pp. 91-137

Chapter 6

Applications of MEMS Gyroscope for Human Gait Analysis

Hongyu Zhao, Sen Qiu, Zhelong Wang, Ning Yang, Jie Li and Jianjun Wang

Abstract

After decades of development, quantitative instruments for human gait analysis have become an important tool for revealing underlying pathologies manifested by gait abnormalities. However, the gold standard instruments (e.g., optical motion capture systems) are commonly expensive and complex while needing expert operation and maintenance and thereby be limited to a small number of specialized gait laboratories. Therefore, in current clinical settings, gait analysis still mainly relies on visual observation and assessment. Due to recent developments in microelectromechanical systems (MEMS) technology, the cost and size of gyroscopes are decreasing, while the accuracy is being improved, which provides an effective way for qualifying gait features. This chapter aims to give a close examination of human gait patterns (normal and abnormal) using gyroscope-based wearable technology. Both healthy subjects and hemiparesis patients participated in the experiment, and experimental results show that foot-mounted gyroscopes could assess gait abnormalities in both temporal and spatial domains. Gait analysis systems constructed of wearable gyroscopes can be more easily used in both clinical and home environments than their gold standard counterparts, which have few requirements for operation, maintenance, and working environment, thereby suggesting a promising future for gait analysis.

Keywords: inertial sensors, inertial measurement units (IMU), gait detection, gait features, gait abnormalities, gait disorders, wearable sensors, body sensor networks (BNS), medical applications

1. Introduction

Gait analysis is the analysis of various aspects of the patterns when we walk or run, which are the most common forms of human legged locomotion, as shown in **Figure 1**. Normal gait is achieved when the multiple body systems function properly and harmoniously, including visual, vestibular, proprioceptive, musculoskeletal, cardiopulmonary, nervous systems, etc. Injury or disease of any system may result in abnormal gait with symptoms and dysfunction of joints and muscles [3–5]. Therefore, gait performance is considered to be an indicator and predictor of overall health and functional status of individuals [6–8]. Gait analysis is an active research area for many medical, clinical, and healthcare applications. The validity and reliability of gait analysis depend strongly on the used measuring instruments. Generally, high-quality gait analysis requires accurate, detailed, and comprehensive spatiotemporal characterization of the actual locomotion pattern.

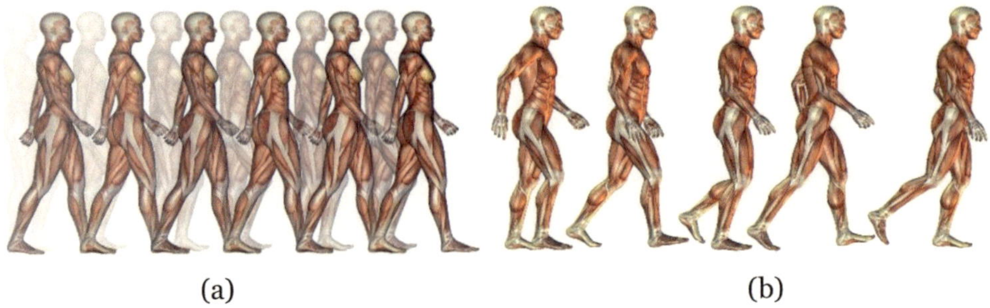

Figure 1.
Human gait. (a) Walking gait [1] and (b) Running gait [2].

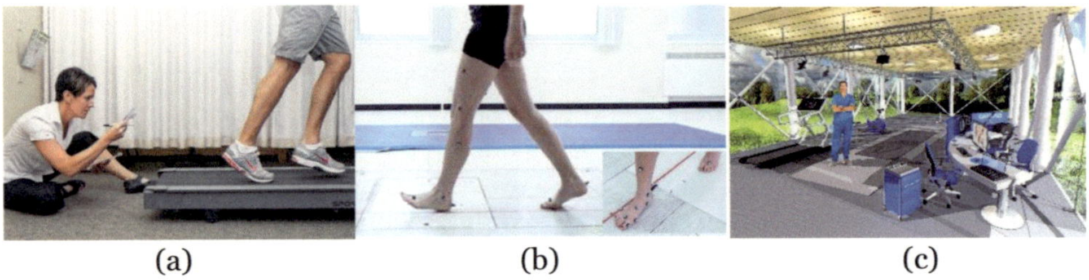

Figure 2.
Commonly used gait analysis methods. (a) Visual gait analysis [9], (b) Vicon systems [10], and (c) BTS GAITLAB [1].

At present, gait analysis in most clinics and health centers is still mainly achieved by patient self-reporting and clinician (physician, nurse, etc.) observation, as shown in **Figure 2(a)**. These subjective and qualitative methods are only suitable for preliminary gait examination. Although some severe gait abnormalities can be visually observed by human eyes, subtle differences might be overlooked without quantitative measurements [11]. With the aid of simple tools like measuring tape, stopwatch, and goniometers, as well as methods allowing leaving footprints on the ground, basic quantitative measures can be derived, such as the number of walked strides/steps, gait cadence, gait speed, stride/step length, stride time, and distance covered. The advantages of the visual observation method and foot-printing method lie in several aspects: (1) they do not require any expensive measuring instruments and complex preparation procedures; (2) they have no special requirements for the working environment; and (3) they can achieve a preliminary gait analysis in a very short time. However, the obtained measures are too limited to assess human gait, as gait is complex and multifactorial in terms of its control mechanisms governed by the neuromuscular system. Besides, the quality of measures is dependent on the observer's experience and the patient's tolerance, especially the inter- and intra-observer variability, which has been shown to significantly influence the disease-specific severity assessment and the subsequent treatment planning.

To provide high-quality quantitative information and objective measurements (some of which might not be measurable with normal clinical examinations) needed for gait analysis, gold standard gait analysis tools have been applied in some specialized centers and clinics, such as optical motion capture systems and force plates. The commonly used such systems are illustrated in **Figure 2(b)** and **(c)**, where the optoelectronic systems capture spatial gait information with infrared cameras tracking the body movement (defined by reflective markers placed on the body), while the force plates provide dynamic gait information by the measuring ground reaction forces (GRFs) based on inverse dynamics. When synchronized with each other, these systems can provide both kinematic and dynamic gait information during walking and running. However, although such systems can achieve high-precision gait analysis, they also have many drawbacks, such as their relatively high cost, long setup time, and complicated operation. Furthermore, they are confined to the restricted area where the systems are deployed and hence affect normal movement of the subjects, which may make the derived information fail to reflect the gait patterns in real-world settings. Generally, people only show their natural gait when they are accustomed to the walking environments.

Electromyography (EMG) systems are another quantitative gait analysis technology commonly used in gait-related applications. Such systems can record the electrical signals generated by skeletal muscles and hence provide insights into the patterns of muscle recruitment and neuromuscular control during walking. They are particularly suitable for investigating gait abnormalities manifested by muscle weakness and spasticity. However, EMG measuring is inconvenient in daily usage, as it requires gel, skin treatment, or smart clothes with embedded textile electrodes, especially for the traditional EMG systems that have intricate wires connecting the electrodes and the signal processor.

For gait analysis, accuracy is not always the only or primary concern, and other relevant concerns include simplicity, accessibility, portability, etc. For example, it might be more meaningful to monitor gait patterns for patients or elders in their daily lives than just a brief examination in a clinic or a gait lab [12]. Therefore, although the optoelectronic, force platform, and EMG systems have been applied to gait analysis in the past decades, they are not pervasive enough, even in specialized centers and clinics, which makes the potential of gait analysis not been fully exploited thus far. In order to make gait analysis more accessible and usable, the use of alternative instruments has been investigated to address the limitations of the gold standard methods, such as inclinometers, goniometers, air pressure sensors, foot switches (or force-sensitive resistors), and inertial sensors. These instruments are more portable, convenient, cost-effective, and easy-to-use, among which inertial sensors are widely considered attractive alternatives. Recent advancements in microelectromechanical systems (MEMS) technology paved a way to develop wearable gait analysis systems constructed of inertial measurement units (IMUs), which have shown remarkable progress in the last two decades. MEMS inertial sensors include gyroscope, accelerometer, as well as a combination of gyroscope, accelerometer, and magnetometer [13, 14]. The commonly used MEMS IMUs in the literature are shown in **Figure 3**.

Notable use of inertial sensors in gait analysis is in providing rich kinematic information about the movement patterns of different body segments. However, there are issues related to the accuracy of the measurements from these low-cost

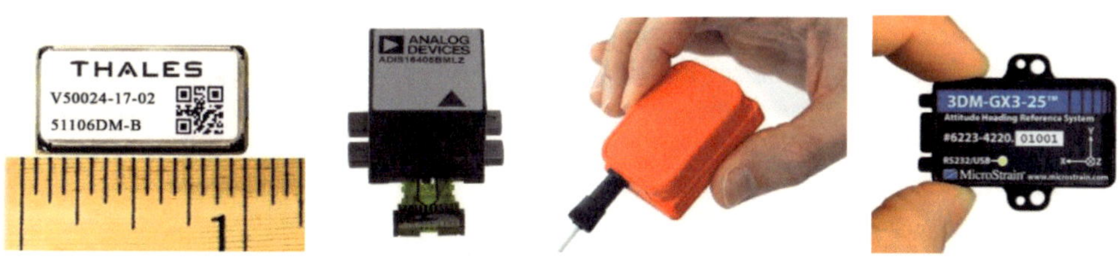

(a) (b) (c) (d)

Figure 3.
Commonly used inertial sensors based on MEMS technology. (a) InterSense IMU, (b) ADI IMU, (c) Xsens IMU, and (d) MicroStrain IMU.

MEMS sensors. The derived angle and position estimates are usually corrupted by varying sensor noises and biases, thereby resulting in the well-known continuously increasing error called drift, i.e., angular or positional deviations away from the ground truth. Many researchers addressed these issues and presented different methods to improve the system accuracy. It should be noted that the costs of MEMS accelerometers are decreasing while the accuracy is being improved, whereas the MEMS low-cost gyroscopes could not achieve the required accuracy for precise long-term positioning applications. Generally, MEMS gyroscopes have large bias drifts that can accumulate several degrees of orientation error during even 1 min. Such large error rates make it difficult to choose reasonably priced gyroscopes for inertial navigation applications, and hence the reliability and accuracy of gyroscopes are questionable [15, 16]. However, for gait analysis applications, gyroscope is the preferred device among the inertial sensors, due to the effects of human locomotion that rotational motion is more pronounced than translational motion. The systems using solely gyroscopes can provide both temporal and spatial gait parameters, whereas most other systems using accelerometers or foot switches are limited to temporal parameters merely. Therefore, the purpose of this chapter is to demonstrate the applications of MEMS gyroscope for human gait analysis.

2. Gait characterization

Generally, pathological gait shows a characteristic pattern with abnormal speed and range of joint movements, such as shortened stance phase, reduced gait cadence, limited extension/flexion, or inversion/eversion ankle movements. Professional physicians could easily recognize gait abnormalities and visually evaluate patients' progress during the physiotherapy treatments; however, quantitative measures allow a detailed description of these abnormalities, which would be desirable for diagnostic and therapeutic use. In this section, the system setup and ankle angles are first described, then the typical modes of dividing a gait cycle are discussed, and finally step lengths that can be provided by foot-mounted gyroscopes are discussed.

2.1 Ankle angles

In biomechanical analysis, kinematic information is a well-established set of gait parameters. To estimate spatiotemporal parameters, wearable gait analysis systems have been discussed in the literature, with two, three, four, or more gyroscopes attached to subject's lower limbs, such as the foot, shank, or thigh. Accurate orientation estimation using gyroscopes has been a major research interest in this field. For wearable systems, a reduction in the number of sensing units is highly desirable, as the system will be more portable, convenient, reliable, cost-effective, and easy-to-use, due to the reduction of total cost and weight, the power consumption and memory requirement, the time and operation needed for system setup, the hindrance to natural movement, etc.

For most types of pedal locomotion achieved by legged motion of human or animals, the intuitive experience is to implement gait analysis by attaching sensors to the feet. As the foot is the part of the lower limb distal to the leg, it functions as the interface between the lower limb and the ground and withstands high static and dynamic stresses that generate strong compression and shearing forces, making the periodic nature and disease symptoms of the foot more obvious than that of other parts of the lower limb. For example, diabetic foot is the distal ankle involvement induced by various causes, mainly because of the interaction of peripheral

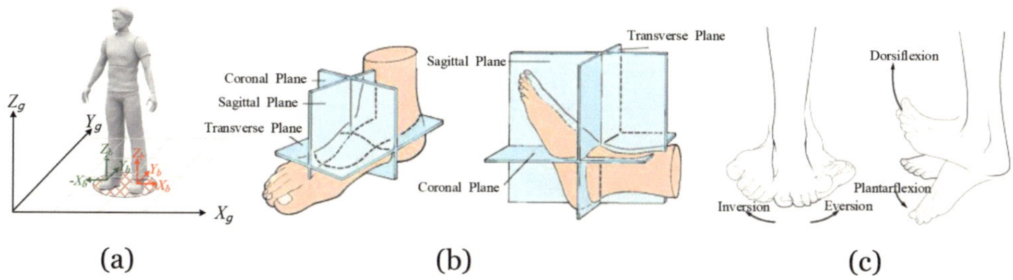

Figure 4.
Dual-sensor configuration for gait analysis and associated ankle angles. (a) Coordinate systems, (b) Foot anatomical planes and (c) Ankle movements.

vasculopathy, neuropathy, and alterations in foot biodynamics [17]. Therefore, the foot is a preferred location of gyroscopes for gait data collection.

A dual-sensor configuration with one sensor on each foot is discussed in this chapter, as illustrated in **Figure 4(a)**, which is supposed to be a promising way for wearable gait analysis. As is seen, two coordinate frames are introduced for gait analysis purpose, which are defined as follows.

- The body coordinate frame (b-frame for short) is parallel to the sensor's axes.

- The global coordinate frame (g-frame for short) is a local east-north-up (ENU) reference frame.

Such system can yield the angles of ankle movements, such as plantarflexion and dorsiflexion movements in the sagittal plane, as well as the inversion and eversion movements in the coronal plane, as shown in **Figure 4(b)** and **(c)**. These movements are described in terms of Euler angles to assess the ankle joint, as ankle rehabilitation includes range of motion training on eversion and inversion as well as plantarflexion and dorsiflexion.

2.2 Gait phases

To analyze gait abnormalities, temporal gait parameters should be estimated first. Terminologically, gait is the movement pattern involved during locomotion, which exhibits periodic patterns termed as gait cycle. Each gait cycle is characterized by a sequence of ordered gait events that occur at specific temporal locations. These events can be detected by using the measurements of wearable MEMS gyroscope. Different researchers pay attention to different gait events according to their specific application requirements. Normally, there are four typical events in one gait cycle, i.e., heel-strike (HS), foot-flat (FF), heel-off (HO), and toe-off (TO), as shown in **Figure 5** identified relative to the right foot and defined in the following way:

1. **HS event**: the heel strikes the ground.

2. **FF event**: the toe touches the ground, and the foot becomes completely flat on the ground.

3. **HO event**: the heel leaves the ground.

4. **TO event**: the toe leaves the ground, and the foot becomes totally in the air.

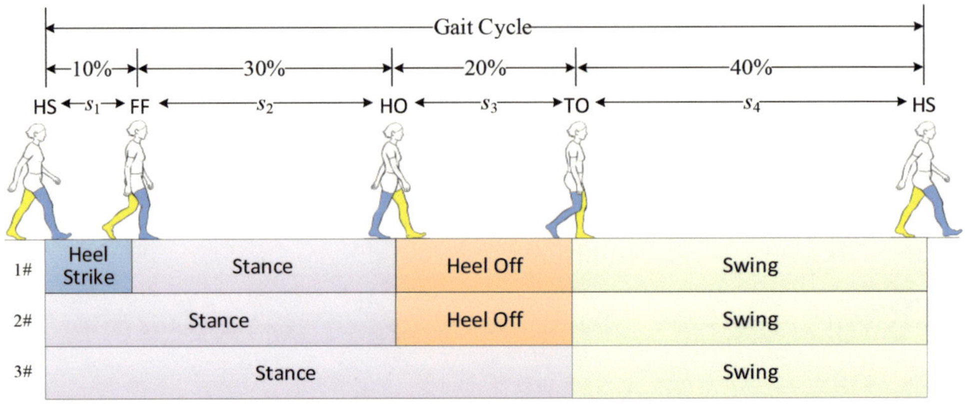

Figure 5.
Typical events and phases in one gait cycle.

Usually, HS event is specified as the beginning of a gait cycle, and a complete gait cycle is defined as the time interval between successive HS events of the same foot. The typical gait events can divide a gait cycle into two to four consecutive time intervals termed as gait phases. When considering more gait events, e.g., mid-stance and mid-swing, more gait phases will be delimited, which is not addressed in this chapter. As shown in **Figure 5**, there are three common modes of gait cycle division. The first mode (1) divides a gait cycle into two phases, i.e., stance and swing, where the stance phase lasts from HS to TO corresponding to about 60% of a gait cycle [18]. The second mode (2) divides a gait cycle into three phases, where the stance phase is delimited by HS and HO constituting about 40% of a gait cycle [19]. The third mode (3) divides a gait cycle into four phases, where the stance phase lasts from FF to HO comprising about 30% of a gait cycle [20].

Obviously, out-of-sequence events are not permitted in normal gait, and hence a breakdown in gait rhythm and bilateral coordination plays a significant role in identifying pathologic gait, e.g., freezing of gait in Parkinson's disease. Besides, for the patients with gait abnormalities, the affected lower limb fails to support the body weight well, which makes the corresponding stance phase short-lasting and results in a highly unstable situation. Monitoring gait cycle distribution in temporal domain has been applied to detect the onset of neurodegenerative diseases and injuries [21].

In this chapter, for demonstration purpose, the mentioned four typical gait events are modeled and identified, and hence a normalized gait cycle is divided into four phases as that in the first division mode (1). In this division, the stance phase is the time interval when the foot is entirely on the ground, the swing phase is the time interval when the foot is entirely in the air, and the two remaining phases are the transition states between stance and swing. Furthermore, as the motions of subject's two feet are strongly coupled with each other, detecting gait events using the measurements of both feet is supposed to obtain more accurate results than just using that of the ipsilateral limb. When the concerned gait events of each foot are correctly detected, the gait cycles will be divided, the gait phases will be delimited, and therefore the temporal gait parameters will be derived accordingly.

2.3 Step lengths

When the gait phases are delimited, the spatial gait parameters can be derived accordingly. The distance-related gait parameters involve step length, step width, and step height, corresponding to the maximum covered distance in the forward, lateral, and vertical directions, respectively, over a step, as shown in **Figure 6**. Among these

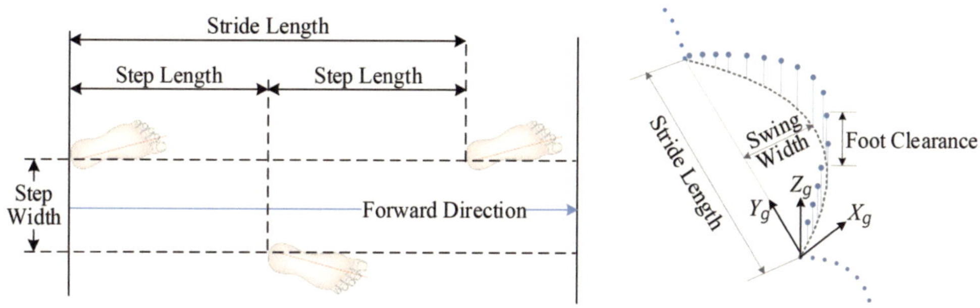

Figure 6.
Distance-related spatial gait parameters.

three parameters, the step length in the sagittal plane needs to be calculated separately for each step of each individual, as it varies considerably due to inter- and intraindividual gait variability. Actually, several factors can account for the phenomenon of gait variability, such as leg length, walking speed, and gait pattern. Step length has different values among the literature data. As reported in [22], the average step length is around 0.75 m for healthy adults walking at their self-selected normal speed of about 1.4 m/s, while, as reported in [23], the average step length varies with gender, which is about 0.79 m for males and 0.66 m for females.

As shown in **Figure 6**, a stride consists of two consecutive steps. Both stride length and step length are meaningful gait parameters to assess gait performance. Gait slowness with reduced step length is a manifestation of diseases affecting walking ability, such as spinal cord injury, stroke, Parkinson's disease, and osteoarticular disorders. Many methods for estimating step length and gait speed have been proposed in the literature, e.g., using a mathematical model. Prior studies have employed a single inverted pendulum model to estimate the step length [24], by using a uniaxial gyroscope. A more sophisticated method presented in [25] employs a double pendulum model comprised of an inverted double pendulum pivoting about the ground during stance and a double pendulum pivoting about the hip during swing. A four-sensor configuration is proposed to deal with the non-pendulum nature of double limb support [26]. Typically, a gait model can be driven by various combinations of direct or indirect gyroscope measurements, with the sensors attached to the subject's shank, thigh, or lower lumbar spine near the body's center of mass (COM), etc. Comparisons between different step length estimators are presented in [27].

Based on different gait models, necessary relations between the step length and various measurable or computable gait variables can be formulated. For the dual-sensor configuration shown in **Figure 4(a)**, a modified gait model was presented in our previous study [28], which is driven by the measurements from foot-mounted gyroscopes solely. In this model, human gait is represented by a single inverted pendulum model of a kneeless biped, taking the anthropometric data specific to each subject's biomechanics into consideration, as shown in **Figure 7**. This model functions as a self-contained step length estimator, which does not simply resort to other ranging technologies based on infrared, RF, or ultrasonic devices that usually use some type of beacon or active badge [29, 30] nor directly double integrate the gravity-compensated translational acceleration over time. The step length S_L can be estimated as the forward distance traversed by the body's COM, during the stance phase of the contralateral rear foot that supports the forward motion of the swing leg.

Therefore, a mathematical model from indirect gyroscope measurements can be adopted to estimate the step length by

$$S_L = L \cdot [\sin(\theta_1) + \sin(\theta_2)] \tag{1}$$

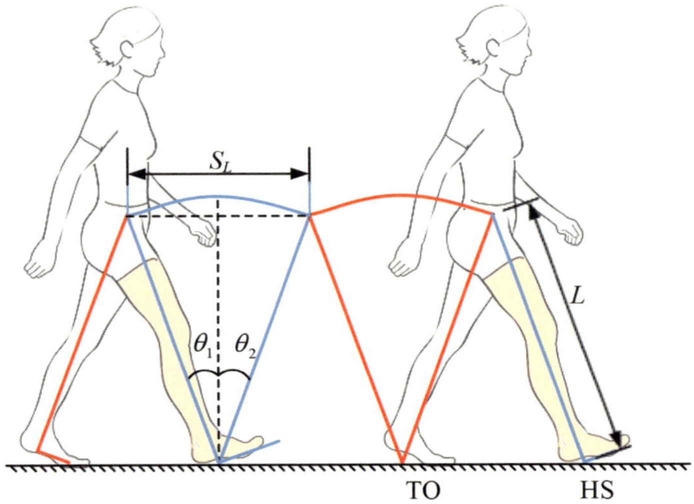

Figure 7.
Kneeless inverted pendulum model of walking gait.

where L denotes the pendulum length related to the subject's height and leg length and θ_1 and θ_2 denote the amplitudes that the pendulum swings away from vertical, which are approximated by the maximum positive and negative rotation angles of foot pitch motion, respectively, and related to the plantarflexion and dorsiflexion angles on the sagittal plane.

3. Gait data acquisition

Gait analysis can be achieved by examining the patterns of sensed data from the measuring instruments. There are two sources of gait data in our study, i.e., inertial data and optoelectronic data. The optoelectronic data are measured by using the Vicon® optical motion capture system from Oxford Metrics Ltd., UK [10], which is used as reference data to provide ground truth for gait analysis algorithms. As illustrated in **Figure 8(a)**, the MEMS inertial sensors and Vicon retroreflective markers are attached to the subject's lower limbs. However, for demonstration purpose, only the measurements from foot-mounted devices are considered in this chapter, as shown in the partial enlarged drawing in **Figure 8(b)**. Two types of inertial

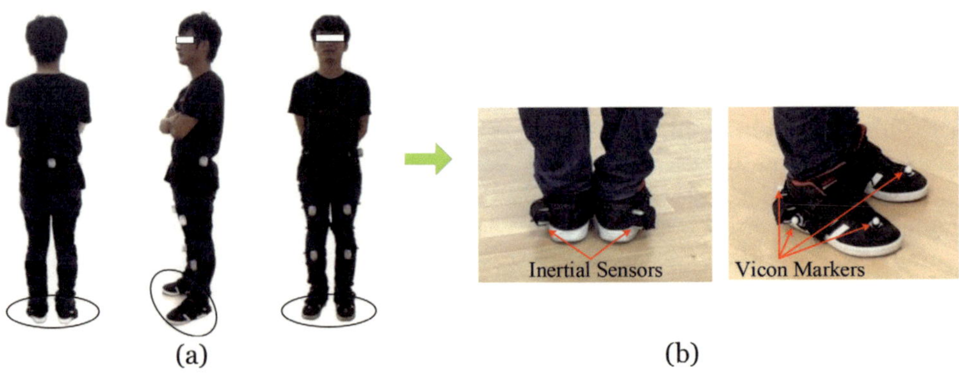

Figure 8.
System setup for gait data acquisition. (a) Sensor placement on lower limbs and (b) Foot-mounted sensors and markers.

sensors are used for gait data collection in our study, i.e., Nano IMU (nIMU) from MEMSense Inc., USA [31], and ADIS16448 *i*Sensor® device from Analog Devices Inc., USA [32], as described below.

3.1 MEMSense IMU

The first type of inertial sensors used in our study is the MEMSense nIMU [13, 33], which is a small-size and low-weight MEMS unit and costs about $1300, as shown in **Figure 9** that illustrates the data acquisition process under normal conditions. Since nIMU is a wired sensor node, when communicating with it, one needs to connect it to a USB interface board first and then connect the USB interface board to a computer for further processing. The software used for acquiring and storing sensed data is the MEMSense IMU Data Console (IDC), which is a console-based, menu-driven application and allows basic display and collection using a specified RS422 protocol.

The nIMU is compensated for temperature sensitivities to bias and scale factor and provides serial outputs including 3D acceleration, 3D angular rate, and 3D magnetic field intensity, with a sampling rate of 150 Hz. The key manufacturer specifications of the gyroscope in nIMU are listed in **Table 1**.

A segment of raw measurements is shown in **Figure 10**. As the IMUs are placed on the subject's feet, the gyroscope measurements feature periodic and repetitive patterns according to the transitions of gait phases. These patterns are helpful for gait analysis, by facilitating the detection of the key gait events and the concerned gait phases correspondingly. Since the feet are exposed to quite extreme dynamics at HS events, it is found that the bandwidth and dynamic ranges of the gyroscope in nIMU are insufficient for optimal gait characterization, as seen in **Figure 10**. These insufficiencies would induce systematic measurement and modeling errors to the system. When testing the sensor for running gait, the achieved tracking results are reasonable but would improve considerably if the gyroscope has sufficient dynamic range, so as to accurately monitor the impact of foot on the ground. According to research in [34], the maximum angular velocity experienced by toe-mounted gyroscopes can

Figure 9.
MEMSense nIMU used for gait data collection.

Mass (g)		20
Size (mm)		45 × 23 × 13
Operating temperature (°C)		0 to +70
Gyroscope	Range (°/s)	±600
	Nonlinearity (% of FS)	±0.1
	Noise (°/s)	0.56 (0.95)
	Bandwidth (Hz)	50

Table 1.
Key specifications of gyroscope in MEMSense nIMU.

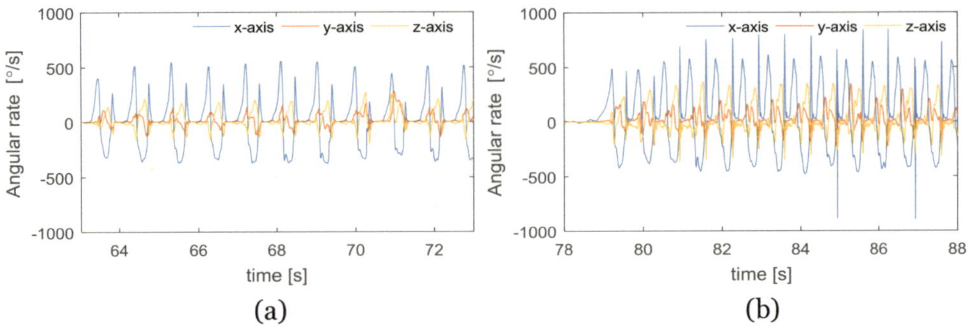

Figure 10.
Raw gyroscope measurements of MEMSense nIMU. (a) Walking at 130 steps/min anf (b) Running at 170 steps/min.

reach 1500°/s during running and 2000°/s during sprinting. This is because the foot attitude changes very rapidly over the gait cycle, especially for the toe motion that exhibits the highest angular velocity. The maximum angular velocities experienced by the heel, ankle, and shin are no higher than 1000°/s during running.

3.2 Analog devices IMU

The other type of inertial sensors used in our study is the ADIS16448 *i*Sensor® device [35–37], which combines industry-leading *i*MEMS® technology with signal conditioning that optimizes dynamic performance and costs about $600. The ADIS16448 is packaged in a module that has a standard connector interface, as illustrated in **Figure 11** that depicts the data acquisition process in a physical therapy and rehabilitation department of a public hospital. The SPI and register structures provide a simple interface for data collection and configuration control. The ADIS16448 has a compatible pinout for systems that currently use other Analog Devices, Inc., IMU products. Each ADIS16448 includes a triaxial gyroscope, a triaxial accelerometer, a triaxial magnetometer, and pressure sensors. The factory calibration characterizes each sensor for sensitivity, bias, and alignment. Thus, each sensor has its own dynamic compensation formulas that provide accurate sensor measurements. The key manufacturer specifications of the gyroscope in ADIS16448 are listed in **Table 2**.

The dimensions of the entire sensing assembly are 4.5 × 3.5 × 2.25 cm, and the sampling rate is 400 Hz. The main components include the ADIS16448 IMU, a printed circuit assembly (PCA) with a microcontroller, a power supply, and a casing enclosing all of the components. The collected data were stored in internal memory first and then transferred to an external computer for further processing. A segment of raw measurements is shown in **Figure 12**. It can be seen that the gyroscope

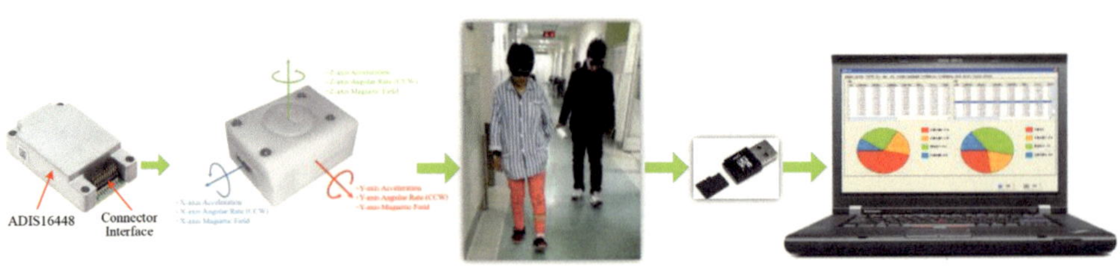

Figure 11.
ADIS16448 iSensor® *device used for gait data collection.*

Mass (g)		15
Size (mm)		24.1 × 37.7 × 10.8
Operating temperature (°C)		−40 to +85
Gyroscope	Range (°/s)	±1000
	Nonlinearity (% of FS)	±0.1
	Noise (°/s)	0.27
	Bandwidth (Hz)	330

Table 2.
Key specifications of gyroscope in ADIS16448 iSensor® device.

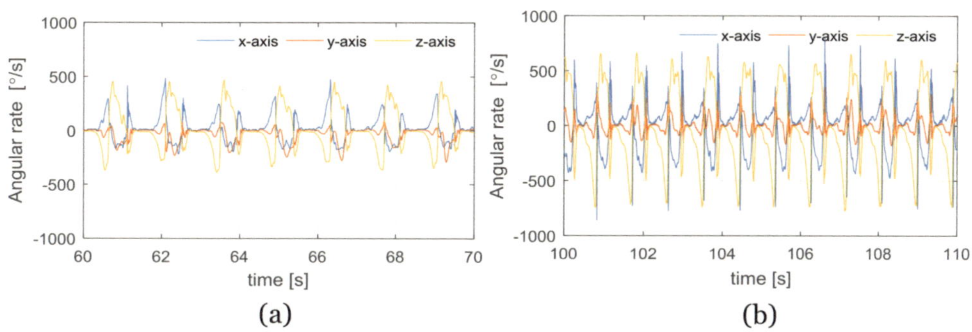

Figure 12.
Raw gyroscope measurements of ADIS16448 iSensor® device. (a) Walking at 3 km/h and (b) Running at 6 km/h.

range of ±1000°/s would be more suitable, as the sensor readings stay within this dynamic range during walking and running at varying speeds.

4. Rule-based gait detection

In this chapter, the raw measurements of accelerometer and gyroscope are compared first, then the gait events are identified by using a rule-based method, and finally the false-detected gait phases are discussed and eliminated.

4.1 Raw inertial measurements

Different methods have been presented for gait detection in the literature [38]. In a sense, gait phases are a function of time and inertial measurements. A segment of raw measurements is shown in **Figure 13**, including specific forces and angular rates of both feet measured by the accelerometer and the gyroscope, respectively, together with the key gait events and their delimited gait phases. Gait detection can be achieved by using a rule-based method from the raw measurements or its magnitude [39], root mean square [40], and moving average [41], which is straightforward and easy to implement. Different detection methods have been compared in [42], and the results suggest that angular rate is more reliable than acceleration for typical walking. As can be seen in **Figure 13**, the angular rates provide more prominent characteristics than the specific forces for gait detection, especially the angular rate around Z-axis in the sagittal plane. Due to the specificity of foot motion, there are at least two possible explanations for this phenomenon:

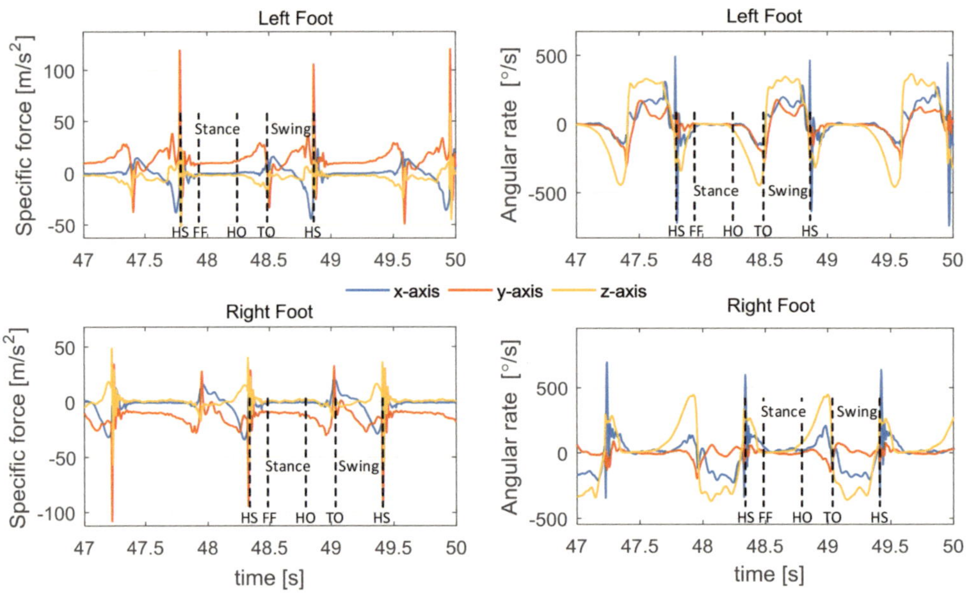

Figure 13.
A segment of raw inertial measurements from both feet.

1. Although the angular rates have large bias, their SNR (signal-to-noise ratio) is higher than that of the specific forces.

2. The specific forces are perturbed by the integrated effects of initial alignment error, gravity disturbance, and accelerometer bias.

4.2 Gait detection with predefined rules

The rules for gait detection from inertial data can be predefined against the ground truth provided by the Vicon system. Generally, three types of rules might be involved in the detection process, i.e., peak detection, flat-zone detection, and zero-crossing detection. Take the stance phase, for example, it is the nature of walking or running locomotion that the foot swings to stance phase in every gait cycle and then exhibits a zero velocity until it swings again. This information can be effectively utilized by a flat-zone detection method to identify the successive stance phases. With careful rule design and parameter selection, the rule-based methods can identify all concerned events from a long inertial data sequence.

For a straight-line walking of 20 m long, the detection results are shown in **Figure 14**. However, as seen in **Figure 13**, the measurements are characterized by some sudden spikes, especially when the HS and TO events occur, which can induce momentary fluctuations in the magnitude or short-term statistics of angular rates and thereby result in false detections of gait phases. In some research, a time heuristic method is applied to the raw detection results to avoid unnecessary influence of the measurement fluctuations, i.e., incorrectly declaring, interrupting, or missing of gait phases. This is achieved by adding a time duration threshold to filter out the gait phases that have a duration shorter than the threshold, as the false gait phases are usually short-lasting [43, 44]. However, as all the thresholds are hand-tuned, they may work well for the gait data that they are derived from, but not apply to each subject's individual gait.

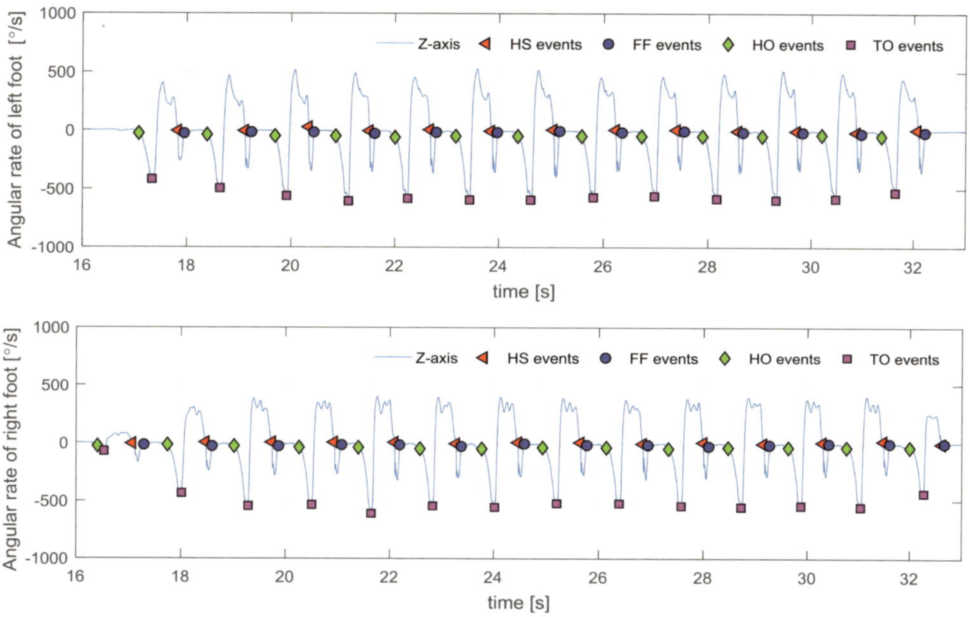

Figure 14.
Results of gait division and gait detection.

4.3 Elimination of false gait phases

As each individual has a unique gait pattern, the percentage of the gait cycle spent in each phase slightly varies between the literature sources. In literature research, it is rarely discussed explicitly how to choose a time duration threshold for eliminating the false gait phases but based on empirical evidence. Therefore, an adaptive time threshold is required to provide a more robust method for gait detection. As done in our previous study, a clustering technique can be used to automatically distinguish the true and false gait phases according to their time durations and yield the time threshold parameter simultaneously [33], as shown in **Figure 15**. In this scenario, since the number of clusters is known, the k-mean or k-median methods can be employed due to their simplicity and efficiency.

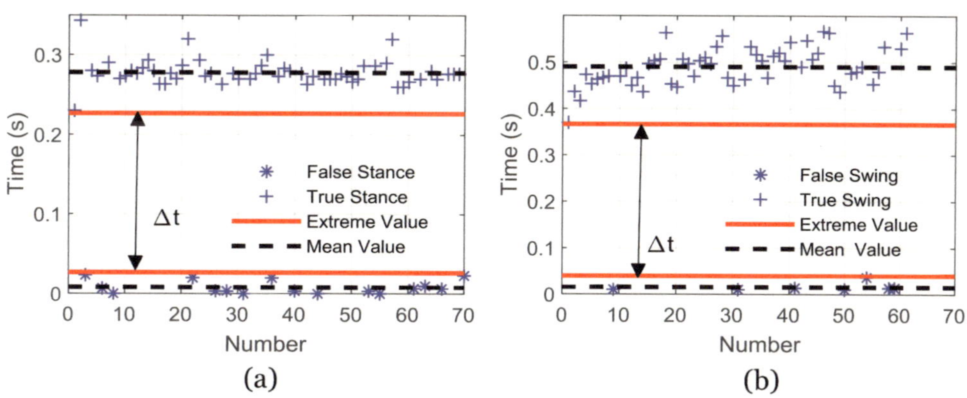

Figure 15.
Binary classification of potential gait phases. (a) Clustering of stance phases and (b) Clustering of swing phases.

Multiple parameters are involved for gait detection, which are interrelated and work together to achieve their goal. The adopted clustering technique tried to tune one of the thresholds automatically (i.e., the time threshold of gait phases) and further facilitate the choice of other thresholds, but careful parameter setting is stilled needed. Generally, rule-based methods rely on careful sensor alignment and a set of thresholds, which are brittle or difficult to implement due to the natural variability of human gait. Moreover, the thresholds are usually hand-tuned and fixed in the whole process regardless of gait changes, and the process of rule designing and threshold tuning itself is frustrating and time-consuming. Furthermore, if new sensors are added to the setup or the sensors are attached to new locations, new detection rules and associated thresholds are required. Therefore, there is a clear need of an adaptive detection method.

5. Machine learning-based gait detection

As mentioned above, gait detection is actually a pattern recognition problem. Hidden Markov models (HMMs) have been widely used for pattern recognition. An HMM-based method was developed for gait detection in children with and without hemiplegia, and the gait events were specified as hidden states [45]. A classifier based on HMM is applied for gait phase detection and discrimination between walking-jogging activities [20]. An HMM was applied to detect the gait phases of children with cerebral palsy [46]. However, HMMs are less suitable for gait data of high dimension. An HMM was adopted to estimate temporal gait parameters with a feature selection and model parametrization system based on genetic algorithms (GAs) [47]. An HMM was presented to detect gait phases with observations provided by a five-layer feed-forward neural network (FNN) [48]. Generally, these hybrid methods have better performance than the pure HMMs when dealing with high-dimensional data. Inspired by the existing methods, an adaptive hybrid method is presented in our previous study [36], by modeling human gait with a left-right HMM and employing a three-layer neural network (NN) to deal with the raw measurements.

5.1 HMM-based gait model

HMM is a statistical model used to represent discrete and stochastic Markov process, in which the states cannot be directly observed. It can be of three types, i.e., ergodic, left-right, or parallel left-right. At each time instant, HMM is in just one state. For gait detection, the gait events or their delimited phases are the hidden states of HMM. Due to the periodic nature of normal foot motion with a sequence of ordered gait events, each state can only transit to itself or the "right" state. Thus, each gait phase can be represented by a unique state in HMM using a left-right model, as shown in **Figure 16**, where a_{ij} is the state transition probability. This process yields a sequence of hidden states and a sequence of corresponding observations. Each HMM state corresponds to a gait phase that begins with the present gait event and lasts until the next event.

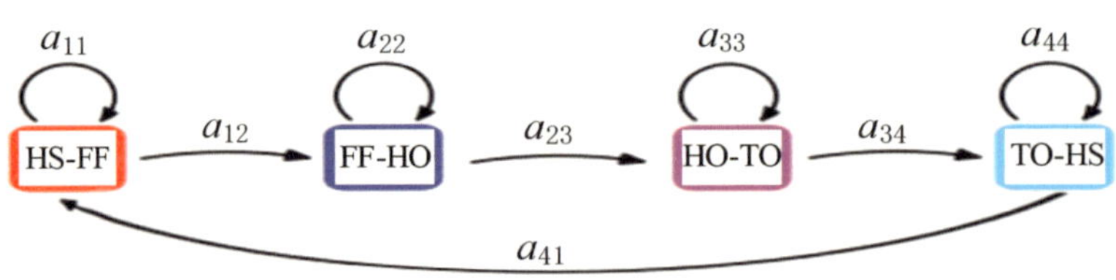

Figure 16.
Left-right HMM with four gait phases.

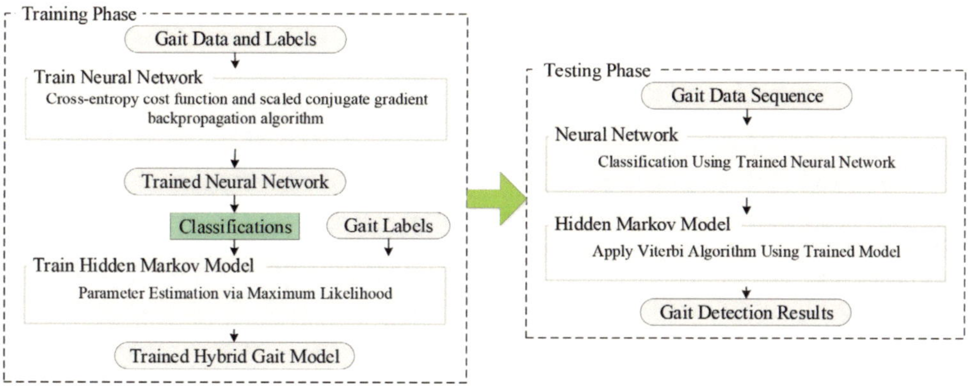

Figure 17.
Framework of the hybrid gait detection method.

5.2 NN-/HMM-based hybrid gait model

Given a sequence of ordered observations and a trained HMM, the Viterbi algorithm can estimate the most likely sequence of hidden states. However, HMMs are generative models, whereas discriminative models are supposed to achieve better classification results. Discriminative models based on machine learning techniques are perceived to be promising alternatives to HMMs [49]. Generally, any machine learning method, such as support vector machine (SVM), k-nearest neighbor (k-NN), and neural network (NN), can be used for gait detection. It is found that NNs can achieve the best trade-off between efficiency, accuracy, and computational complexity. The NNs can learn nonlinear combinations of inputs automatically, and a three-layer network can approximate any multivariate polynomial function [50]. However, the pure NNs have been limited to process inputs in isolation.

To take advantage of both NN and HMM methods for gait detection, one intuitive way is to combine them together in a hybrid manner [48]. The NN can process the gyroscope measurements first and provide observations for HMM with its classifications. Each input of NN is formed by using a sliding window approach, and hence it might be of high dimension. The HMM can model the sequential property of human gait and complement the NN by providing contextual information. **Figure 17** shows the framework of training and testing procedures of this hybrid detection method. Although the NN-/HMM-based hybrid method is computationally complex for training, it is computationally efficient at runtime. It requires no careful sensor alignment or parameter adjustment and generalizes well to new subjects, new gaits, new sensors, and new sensor locations [51].

6. Gait analysis experiment

Usually, pathological gait exhibits a characteristic gait pattern with limited range and velocity, such as shortened stance phase and step length, reduced gait cadence and gait velocity, and diminished extension-flexion movement. The outputs of wearable gait analysis system are of great use for a close examination of human gait, which allow a rapid and accurate quantification of these abnormalities. In this section, the setup and results of the experiments are first presented, then some discussions on the experimental results are made, and finally the capability of IMU-based gait analysis system for tracking the rehabilitation process is verified.

6.1 Experiment setup

Patients during the course of their rehabilitation were recruited as volunteers in our study. For comparison purpose, young healthy subjects were also recruited as volunteers to participate in the study. Prior to each trial, the subjects were asked to stand still for a few seconds to perform initial alignment of the system [52]. During each trial, the patients were instructed to walk at their comfortable speed along a straight-line path about 10 m long, which is along a hospital corridor and free of obstacles. All patients were asked to perform two consecutive walks in forward and backward directions, respectively, and return at the starting position at the end of each trial. For the healthy subjects, the experiment was performed with the same procedure, except that four consecutive walks were performed and the predefined path was 20 m long on a flat floor of a modern office building.

6.2 Experiment results

6.2.1 Single trial

When the gait events are correctly detected, the spatiotemporal gait parameters can be extracted, such as gait phase duration, gait cycle distribution, foot angle, stride length, and gait speed, as shown in **Figure 18** for a single trial.

As is seen in **Figure 18**, a gait cycle is divided into four successive phases, which are defined as follows:

- HS (HS-FF): the phase lasting from HS event to FF event
- ST (FF-HO): the stance phase
- HO (HO-TO): the phase lasting from HO event to TO event
- SW (HO-HS): the swing phase

Hemiparesis can lead to unilateral paresis, i.e., weakness of one side of the body. Compared with the normal gait of healthy subjects, several conclusions can be drawn for the pathological gait of patients from the results shown in **Figure 18**, some of which are as follows:

1. The patient exhibits a reduced gait cadence with longer gait cycle and an irregular and asymmetric gait pattern.

2. The patient is affected on the left side, as the stance phases of the left foot are shorter than that of the right side, while the opposite is true for the swing phases, which is in turn due to the affected lower limb that cannot support the body weight well alone and creates a highly unstable situation.

3. The patient exhibits a significantly diminished extension-flexion foot movement, especially for his left foot, where the shortened HS phases are manifestations of insufficient foot dorsiflexion during the swing phase.

4. The patient exhibits a shortened stride length, as although the covered distance of the patient is half that of the healthy subject, the numbers of strides taken were almost the same.

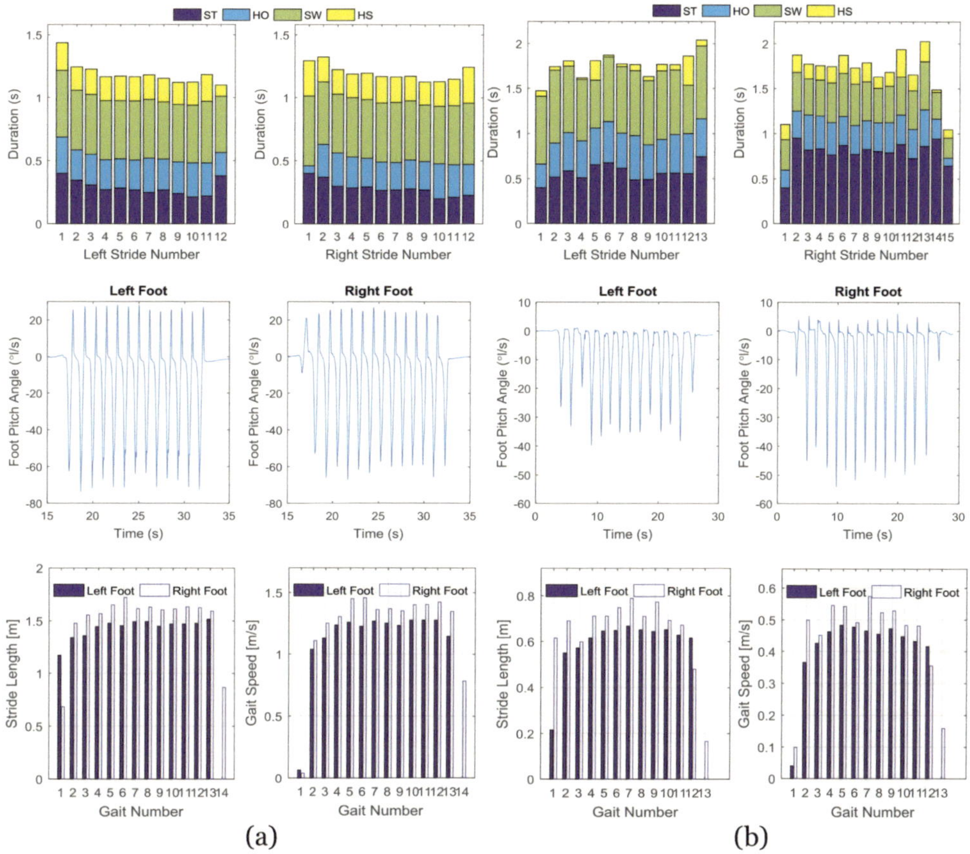

Figure 18.
Estimated spatiotemporal gait parameters. (a) Healthy subject and (b) Hemiparesis patient.

5. The patient's gait speeds of both feet are clearly reduced, and the stride lengths of both lower limbs are greatly shortened, which supports the hypothesis that the healthy side is influenced by the affected side, and the patient even has no asymptomatic side due to the so-called compensatory responses.

As discussed above, insufficient dorsiflexion of foot motion means that the patient is not capable of lifting the toe adequately during the swing phase, thereby resulting in a quick translation from swing to stance. This disorder could not only yield abnormal proportions of gait phases, affecting the gait symmetry and gait regularity, but also be dangerous to patients for being a high risk of fall as it alters the load distribution.

6.2.2 Multiple trials

More trials were performed for a rich data to increase the variability of gait patterns. For demonstration purpose, the average values and standard deviations of the durations of each gait phase and their relative percentages in each gait cycle are calculated for each concerned gait phase of all the trials, as shown in **Figures 19** and **20**. The result of multiple trials further confirms the conclusions made from that of the single trial.

6.3 Rehabilitation process evaluation

To verify the ability of IMU-based gait analysis system for evaluating the rehabilitation process, a patient's gait was measured once a week for 1 month. The

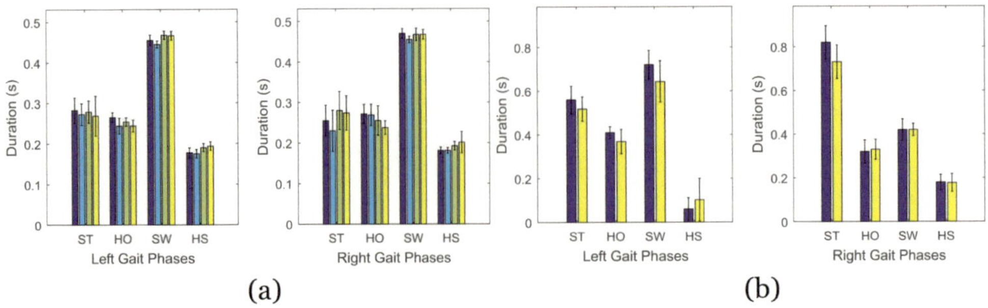

Figure 19.
Durations of the gait phases over multiple trials. (a) Healthy subject and (b) Hemiparesis patient.

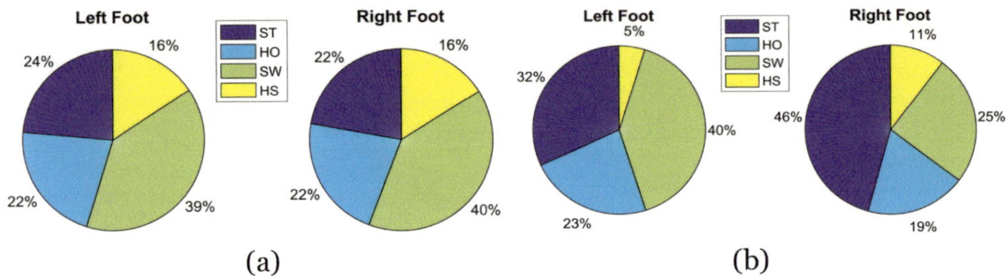

Figure 20.
Pie charts of the gait phases. (a) Healthy subject and (b) Hemiparesis patient.

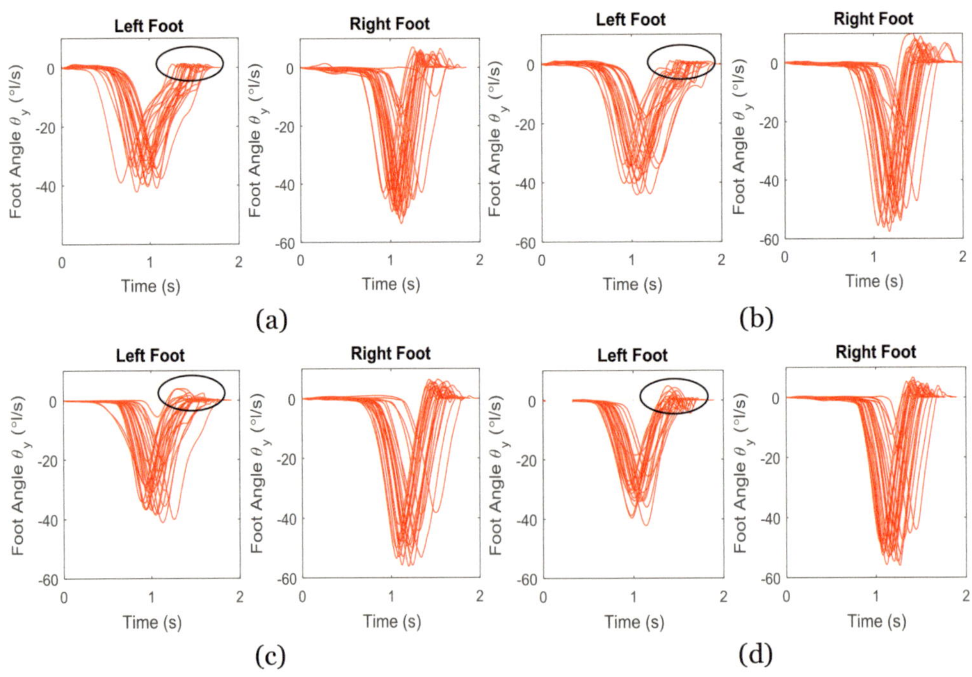

Figure 21.
Aligned extension-flexion angles of foot during rehabilitation. (a) First week, (b) Second week, (c) Third week and (d) Fourth week.

estimated foot angles of extension-flexion movement are shown in **Figure 21**, in which each region of maximum foot dorsiflexion angles is marked by an ellipse. For a better comparison, the angles over successive gait cycles are segmented and aligned along the time axis with the same starting point.

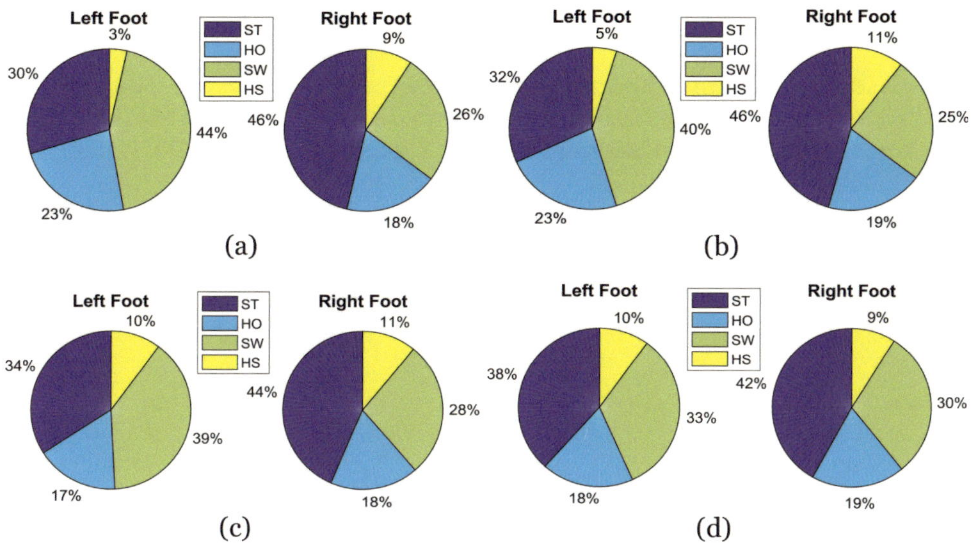

Figure 22.
Averaged percentages of gait phases during rehabilitation. (a) First week, (b) Second week, (c) Third week and (d) Fourth week.

As expected, the dorsiflexion range of the affected foot increased gradually as the treatment proceeded. If not, healthcare professionals may need to modify their treatment plans. Meanwhile, the gait symmetry of patient was also improved, as shown in **Figure 22** in terms of the percentages of the gait cycles spent in each concerned phase.

The conclusions drawn from **Figures 21** and **22** are consistent with that drawn in Sections 6.1 and 6.2. Based on the quantified spatiotemporal gait parameters that are provided by gyroscope-based foot-mounted gait analysis system, gait performance can be characterized to assess the rehabilitation process of patients with gait abnormalities, which is useful for deciding appropriate medical intervention.

7. Conclusions

Quantification of human gait via wearable inertial sensors has been attracting increasing interests in recent years, ranging from aiding pathologic diagnosis, choosing appropriate therapy, evaluating treatment efficacy, and assessing rehabilitation outcomes to monitoring gait degradation, predicting fall risks, and preventing elderly falls. This chapter demonstrated that gait analysis system constructed of foot-mounted MEMS gyroscopes could provide a promising way for estimating spatiotemporal gait parameters and has various potential uses in future research and clinical applications. Such systems are not only convenient for clinical diagnosis and treatment use but also can continuously monitor gait changes in nonclinical settings, thus providing seamless gait analysis from clinical to real-world settings.

However, although wearable technologies are regarded as solutions to create a more effective, convenient, and economical gait analysis technology, the potential of gait analysis has not been fully exploited thus far. There is still a great deal to do for its pervasive use. In future work, more gait parameters will be closely examined in the spatiotemporal domain, to conduct a thorough examination of person's pathological gait. Furthermore, reasonable indexes will be explored to evaluate the gait performance as fully as possible, and some nonlinear analysis techniques will be utilized to provide insight into the neuromuscular control processes that govern human locomotion.

Acknowledgements

This work was jointly supported by the China Postdoctoral Science Foundation no. 2017M621131, National Natural Science Foundation of China no. 61473058, and Fundamental Research Funds for the Central Universities no. DUT18RC(4)036 and DUT16RC(3)015.

Author details

Hongyu Zhao[1,2*], Sen Qiu[2], Zhelong Wang[2], Ning Yang[2], Jie Li[2] and Jianjun Wang[3]

1 Key Laboratory of Intelligent Control and Optimization for Industrial Equipment of Ministry of Education, Dalian University of Technology, Dalian, China

2 School of Control Science and Engineering, Dalian University of Technology, Dalian, China

3 Beijing Institute of Spacecraft System Engineering, Beijing, China

*Address all correspondence to: zhaohy@dlut.edu.cn

References

[1] Bioengineering B. BTS GAITLAB: Integrated Gait Analysis Systems [Internet]. 2019. Available from: https://www.btsbioengineering.com/products/bts-gaitlab-gait-analysis/ [Accessed: 22 March 2019]

[2] Physiotherapy S. Running and gait analysis [Internet]. 2019. Available from: https://southfieldsphysio.co.uk/index.php/services/running-gait-analysis [Accessed: 22 March 2019]

[3] Paramanandam V, Lizarraga KJ, Soh D, Algarni M, Rohani M, Fasano A. Unusual gait disorders: A phenomenological approach and classification. Expert Review of Neurotherapeutics. 2019;**19**(2):119-132

[4] Qiu S, Liu L, Zhao H, Wang Z, Jiang Y. MEMS inertial sensors based gait analysis for rehabilitation assessment via multi-sensor fusion. Micromachines. 2018;**9**(9):442

[5] Qiu S, Wang Z, Zhao H, Liu L, Jiang Y. Using body-worn sensors for preliminary rehabilitation assessment in stroke victims with gait impairment. IEEE Access. 2018;**6**:31249-31258

[6] Morris R, Hickey A, Del Din S, Godfrey A, Lord S, Rochester L. A model of free-living gait: A factor analysis in Parkinson's disease. Gait and Posture. 2017;**52**:68-71

[7] Zhao H, Wang Z, Qiu S, Ning Y, Shen Y. Examination of Gait Disorders in Hemiparesis Patients using Foot-Mounted Inertial Sensors. In: The 6th International Conference on Communications, Signal Processing, and Systems (CSPS '17); 14-16 July 2017; Harbin, China

[8] Zhao H, Wang Z, Qiu S, Shen Y, Wang J. IMU-based Gait Analysis for Rehabilitation Assessment of Patients with Gait Disorders. In: The 4th International Conference on Systems and Informatics (ICSAI '17); 11-13 November 2017; Hangzhou, China

[9] Therapy OP. Gait Analysis [Internet]. 2019. Available from: http://onpoint-pt.com/services/ [Accessed: 26 March 2019]

[10] Vicon. Motion Capture for Life Science [Internet]. 2019. Available from: https://www.vicon.com/motion-capture/life-sciences [Accessed: 26 March 2019]

[11] Chen S, Lach J, Lo B, Yang GZ. Towards pervasive gait analysis for medicine with wearable sensors: A systematic review. IEEE Journal of Biomedical and Health Informatics. 2016;**20**(6):1521-1537

[12] Qiu S, Wang Z, Zhao H, Liu L, Li J, Jiang Y, et al. Body sensor network based robust gait analysis: Toward clinical and at home use. IEEE Sensors Journal. 2019

[13] Zhao H, Wang Z, Shang H, Hu W, Gao Q. A time-controllable Allan variance method for MEMS IMU. Industrial Robot: An International Journal. 2013;**40**(2):111-120

[14] Choe N, Zhao H, Qiu S, So Y. A sensor-to-segment calibration method for motion capture system based on low cost MIMU. Measurement. 2019;**131**:490-500

[15] Zhao H, Wang Z, Gao Q, Hassan MM, Alelaiwi A. Smooth estimation of human foot motion for zero-velocity-update-aided inertial pedestrian navigation system. Sensor Review. 2015;**35**(4):389-400

[16] Qiu S, Wang Z, Zhao H, Qin K, Li Z, Hu H. Inertial/magnetic sensors based pedestrian dead reckoning by means of multi-sensor fusion. Information Fusion. 2018;**39**:108-119

[17] Zavala AV. Vascular and neuropathic foot. In: Cohen Sabban E, Puchulu F, Cusi K, editors. Dermatology and Diabetes. Cham: Springer; 2018. pp. 225-241

[18] Feliz R, Zalama E, Gómez-Bermejo JG. Pedestrian tracking using inertial sensors. Journal of Physical Agents. 2009;**3**(1):35-42

[19] Kotiadis D, Hermens HJ, Veltink PH. Inertial gait phase detection for control of a drop foot stimulator: Inertial sensing for gait phase detection. Medical Engineering and Physics. 2010;**32**(4):287-297

[20] Mannini A, Sabatini AM. Gait phase detection and discrimination between walking–jogging activities using hidden Markov models applied to foot motion data from a gyroscope. Gait and Posture. 2012;**36**(4):657-661

[21] Gouwanda D, Gopalai AA, Khoo BH. A low cost alternative to monitor human gait temporal parameters–wearable wireless gyroscope. IEEE Sensors Journal. 2016;**16**(24):9029-9035

[22] Sessoms PH. Step by step: A study of step length in able-bodied persons, race walkers, and persons with amputation. Dissertation Abstracts International: Section B: The Sciences and Engineering. 2009;**69**:6970

[23] Yamaguchi T, Hatanaka S, Hokkirigawa K. Effect of step length and walking speed on traction coefficient and slip between shoe sole and walkway. Tribology Online. 2008;**3**(2):59-64

[24] Miyazaki S. Long-term unrestrained measurement of stride length and walking velocity utilizing a piezoelectric gyroscope. IEEE Transactions on Biomedical Engineering. 1997;**44**(8):753-759

[25] Aminian K, Najafi B, Büla C, Leyvraz PF, Robert P. Spatio-temporal parameters of gait measured by an ambulatory system using miniature gyroscopes. Journal of Biomechanics. 2002;**35**(5):689-699

[26] Allseits E, Agrawal V, Lučarević J, Gailey R, Gaunaurd I, Bennett C. A practical step length algorithm using lower limb angular velocities. Journal of Biomechanics. 2018;**66**:137-144

[27] Jahn J, Batzer U, Seitz J, Patino-Studencka L, Boronat JG. Comparison and evaluation of acceleration based step length estimators for handheld devices. In: International Conference on Indoor Positioning and Indoor Navigation (IPIN '10); 15-17 September 2010; Zürich, Switzerland

[28] Zhao H, Wang Z, Qiu S, Shen Y, Zhang L, Tang K, et al. Heading drift reduction for foot-mounted inertial navigation system via multi-sensor fusion and dual-gait analysis. IEEE Sensors Journal. 2019

[29] Brand TJ, Phillips RE. Foot-to-foot range measurement as an aid to personal navigation. In: Proceedings of the 59th Annual Meeting of The Institute of Navigation and CIGTF 22nd Guidance Test Symposium; 23-25 June 2003; Albuquerque, NM, USA

[30] Qi Y, Boon SC, Erry G, Kay-Soon L, Rijil T. Estimation of spatial-temporal gait parameters using a low-cost ultrasonic motion analysis system. Sensors. 2014;**14**(8):15434-15457

[31] MEMSense. NANO IMU product specification and user's guide [Internet]. 2019. Available from: https://www.memsense.com/products/legacy/nano [Accessed: 30 March 2019]

[32] Analog Devices. ADIS16448 [Internet]. 2019. Available from: http://www.analog.com/en/products/sensors-mems/inertial-measurement-units/adis16448.html [Accessed: 15 October 2018]

[33] Wang Z, Zhao H, Qiu S, Gao Q. Stance-phase detection for ZUPT-aided foot-mounted pedestrian navigation system. IEEE/ASME Transactions on Mechatronics. 2015;**20**(6):3170-3181

[34] Bancroft J, Lachapelle G. Performance of Pedestrian Navigation Systems as a Function of Sensor Location. In: Proc NATO Symposium Navigation Sensors and Systems in GNSS Denied Environments; 8-9 October 2012; Izmir, Turkey

[35] Wang J, Wang Z, Zhao H, Qiu S, Li J. Using wearable sensors to capture human posture for lumbar movement in competitive swimming. IEEE Transactions on Human-Machine Systems. 2019;**49**(2):194-205

[36] Zhao H, Wang Z, Qiu S, Wang J, Xu F, Wang Z, et al. Adaptive gait detection based on foot-mounted inertial sensors and multi-sensor fusion. Information Fusion. 2019;**52**:157-166

[37] li J, Wang Z, Wang J, Zhao H, Qiu S, Yang N, et al. Inertial sensor-based analysis of equestrian sports between beginner and professional riders under different horse gaits. IEEE Transactions on Instrumentation and Measurement. 2018;**67**(11):2692-2704

[38] Taborri J, Palermo E, Rossi S, Cappa P. Gait partitioning methods: A systematic review. Sensors. 2016;**16**(1):20

[39] Fischer C, Sukumar PT, Hazas M. Tutorial: Implementing a pedestrian tracker using inertial sensors. IEEE Pervasive Computing. 2013;**12**(2):17-27

[40] Strömbäck P, Rantakokko J, Wirkander SL, Alexandersson M, Fors I, Skog I, et al. Foot-mounted inertial navigation and cooperative sensor fusion for indoor positioning. In: Proceedings of the International Technical Meeting of the Institute of Navigation (ITM '10); 25-27 January 2010; San Diego, CA

[41] Abdulrahim K, Hide C, Moore H, Hill C. Integrating low cost IMU with building heading in indoor pedestrian navigation. Journal of Global Positioning Systems. 2011;**10**(1):30-38

[42] Skog I, Händel P, Nilsson JO, Rantakokko J. Zero-velocity detection—An algorithm evaluation. IEEE Transactions on Biomedical Engineering. 2010;**57**(11):2657-2666

[43] Yun XP, Calusdian J, Bachmann ER, McGhee RB. Estimation of human foot motion during normal walking using inertial and magnetic sensor measurements. IEEE Transactions on Instrumentation and Measurement. 2012;**61**(7):2059-2072

[44] Godha S, Lachapelle G. Foot mounted inertial system for pedestrian navigation. Measurement Science and Technology. 2008;**19**(7):1-9

[45] Abaid N, Cappa P, Palermo E, Petrarca M, Porfiri M. Gait detection in children with and without hemiplegia using single-axis wearable gyroscopes. PLoS ONE. 2013;**8**(9):e73152

[46] Taborri J, Scalona E, Palermo E, Rossi S, Cappa P. Validation of inter-subject training for hidden Markov models applied to gait phase detection in children with cerebral palsy. Sensors. 2015;**15**(9):24514-24529

[47] Guenterberg E, Yang AY, Ghasemzadeh H, Jafari R, Bajcsy R, Sastry SS. A method for extracting temporal parameters based on hidden Markov models in body sensor networks with inertial sensors. IEEE Transactions on Information Technology in Biomedicine. 2009;**13**(6):1019-1030

[48] Evans RL, Arvind D. Detection of gait phases using orient specks for mobile clinical gait analysis. In: The 11th

International Conference on Wearable and Implantable Body Sensor Networks; 16-19 June 2014; Zurich, Switzerland

[49] Ogiela MR, Jain LC. Computational Intelligence Paradigms in Advanced Pattern Classification. Berlin Heidelberg: Springer; 2012

[50] Hecht-Nielsen R. Theory of the backpropagation neural network. In: Wechsler H, editor. Neural Networks for Perception. Academic Press; 1992. pp. 65-93. https://www.sciencedirect.com/book/9780127412528/neural-networks-for-perception#book-description

[51] Zhao H, Wang Z, Qiu S, Li J, Gao F, Wang J. Evaluation of inertial sensor configurations for wearable gait analysis systems. In: The 20th IEEE/ACIS International Conference on Software Engineering, Artificial Intelligence, Networking and Parallel/Distributed Computing (SNPD '19); 8-10 July 2019; Toyama, Japan

[52] Zhao H, Shang H, Wang Z, Jiang M. Comparison of initial alignment methods for SINS. In: World Congress on Intelligent Control and Automation (WCICA '11); 21-25 June 2011; Taipei, Taiwan